Safety Fundamentals and Best Practices in Construction Industry

Safety Fundamentals and Best Practices in Construction Industry

Dr. Pedro P. Marfa, PhD, MBA, BSCE, OSH Consultant

Copyright © 2016 by Pedro P. Marfa.

Library of Congress Control Number: 2016909084
ISBN: Hardcover   978-1-5144-9670-1
         Softcover   978-1-5144-9672-5
         eBook   978-1-5144-9671-8

All rights reserved. No part of this book may be reproduced or transmitted in any form or by any means, electronic or mechanical, including photocopying, recording, or by any information storage and retrieval system, without permission in writing from the copyright owner.

Any people depicted in stock imagery provided by Thinkstock are models, and such images are being used for illustrative purposes only.
Certain stock imagery © Thinkstock.

Print information available on the last page.

Rev. date: 06/08/2016

To order additional copies of this book, contact:
Xlibris
1-800-455-039
www.Xlibris.com.au
Orders@Xlibris.com.au
721210

Contents

Acknowledgement ix
Message from the Author xi
Introduction xiii

Chapter 1 – Initial Project Planning. 1
    Administrative Framework 2
    Preliminary Safety Requirements 3
    Standard Operating Procedures 4
    Clients' Internal Safety Requirements 6
    Employees' Involvement 8

**Chapter 2 – Overview of Occupational Health
               and Safety Management**. 10
    General Principles 11
    Implementation of Occupational Health and
        Safety Standards. 12
    Administrative Safety Management 14
    Qualifications of Safety Personnel and Training
        Certification. 15
    Employee Safety Orientation 18
    Preliminary Incident Report. 19
    Safety Meetings. 20
    Safety Inspections 21
    Housekeeping 22
    Contractor Laydown and Fabrication Yard 23

**Chapter 3 – Construction Management
               Assignment of Responsibilities** 26
    Management Commitment. 27

    Health and Safety Policy........................... 28
    Employer's Responsibilities 30
    Project Manager Responsibilities 31
    Safety Manager Responsibilities..................... 33
    Project Engineer Responsibilities 33
    Site Supervisor/Foreman Responsibilities 34
    Workers' Duties and Responsibilities 35
    Workers' Cooperation 36

Chapter 4 – Organizational Structure...................... 38
    Key Principle of HSE 38
    Structured Safety Organizational Chart................ 40
    Centralized Management.......................... 42
    Decentralized Management 43

Chapter 5 – Lessons Learned from Unsafe Acts
    and Conditions................................ 45
    Unsafe Acts/Conditions of Mobile Scaffolds 47
    Recommendations 48
    Unsafe Conditions of Free-Standing Scaffolds 49
    Recommendations 51
    Unsafe Acts of Working at Heights 51
    Recommendations 52
    Unsafe Acts of Workers Standing in Modified Man Basket 53
    Recommendations 55
    Unsafe Acts of Using a Stepladder 55
    Recommendations 56
    Unsafe Acts/Conditions in Excavation 57
    Recommendations 59
    Unsafe Conditions of Power Hand Tools 60
    Recommendations 61

Chapter 6 – Hierarchy of Hazards Control 63
    Elimination 63
    Substitution 64
    Engineered Controls 64
    Temporary Working Platforms 65
    Administrative Controls 68
    Personal Protective Equipment (PPE) 68

Chapter 7 – How to Conduct Safety Meetings Effectively 72
    Safety Topic 1: Eye Protection 73
    Safety Topic 2: Hearing Protection 75
    Safety Topic 3: Hand Protection 77
    Safety Topic 4: Fall Protection 78
    Safety Topic 5: Stepladders 80
    Safety Topic 6: Extension Ladders 82
    Safety Topic 7: Electrical Safety 84
    Safety Topic 8: Trenching and Excavation Inspection 87
    Safety Topic 9: Confined Spaces – Dangerous
        Atmosphere 89
    Safety Topic 10: Housekeeping 92

Chapter 8 – Developing a Hazard Identification Plan (HIP) 95
    How to Develop a Hazard Identification Plan (HIP) 97
    Trenching and Excavation 98
    Rebar Works 101
    Painting in Confined Space 102
    Welding Work 104
    Heavy Equipment Operation 106

Chapter 9 – Developing a Method Statement 108
    Purpose 110

  Scope 111
  References 111
  Methodology 111
  General Safety Requirements 113

Chapter 10 – Steps in Developing a Job Safety
     Analysis (JSA) 115
  Sequence of Basic Steps 118

Appendix K – Hazard Identification Plan (HIP) Form 123
Appendix L – Job Safety Analysis (JSA) Form 124
References 125
Special Thanks 127
About the Author 129

Acknowledgement

It took some time to write down my twenty-five years of experience in the field of safety in oil and gas facility projects. This was made possible through the assistance of my fellow officers in the Philippine Society of Safety Practitioners, who shared their ideas and filled this book with best safety practices applied in the construction industry.

To all my friends, officemates, and colleagues in the society, please accept my special thanks to all of you for your contributions and encouragement to complete my book. With all of your advice and recommendations that has been given to me, I've got the opportunity to transform my field of experience into this book.

To William "Robert" Crumpton IV, CMIOSH, CSP, CHMM; Engr. Jesus G. Pedines Jr., author of *Think and Become Safety Practitioner*; Engr. Peter Q. Panganiban, Safety Advisor; and Alan K. D. Abellana, PME, who have offered their precious time to read through my book and provide excellent recommendations, contributions, and suggestions to make this book more beneficial to the reader. I also acknowledge Ms. Charisse Marnelli M. Gabunada who has thoroughly reviewed my manuscript. Your combined professional contributions will help everyone learn step-by-step guidelines on safety fundamentals as required in the construction industry.

Special thanks to my wife, Verna O. Marfa and kids, Sheryl Anne; Rico Lyndon; and Carlo Jim, who were always with me to provide strength and guidance to make this book a reality. Your untiring support of my passion of sharing my knowledge of safety for

Dr. Pedro P. Marfa, PhD, MBA, BSCE, OSH Consultant

professionals working in the construction industry around the globe will help them promote the protection of the people and environment at the construction site.

Sincerely yours,
Pedro P. Marfa, PhD, MBA, BSCE, OSH Consultant

Message from the Author

I would like to dedicate this book to you in hopes of supplementing your knowledge about health and safety requirements in the construction industry while you are in a journey of your career. During more than twenty-five years of experience in construction at petrochemical plants, I have learned that a lot of safety standards, regulation, and procedures are available from the client's safety department or even in your office shelves that could probably kept beside you for a long time. All these standards are a good reference to perform your daily work, supporting projects in a timely manner, with a view to completion with quality and safety. However, few people are reading the safety requirements of the projects and trying to implement them in the construction sites.

Theoretically, most of the personnel reading the safety standards are safety managers, safety supervisors and safety officers. As a result, most have failed to attain full safety compliance with their projects due to a delusion that safety personnel are non-productive; that nobody would listen to the recommendation of best safety practices, which would delay the job; and that overhead costs for the project would rise. Contrary to the spirit of applicable safety requirements, site supervision instead often cut corners, doing unsafe acts while the safety officer is not on site. This unsafe behaviour exposes workers to potential hazards in the workplace, and accidents can happen in an instant. Management staff must realize that these incidents may be prevented if safety procedures are followed. Above all, they must realize this ahead of time, or they do not help workers' protection or prevent the need to recover loss or damages.

Dr. Pedro P. Marfa, PhD, MBA, BSCE, OSH Consultant

I am writing this book because I want to reach out to project engineers, site supervisors, and safety officers to let them know that the safety standards, regulations, and procedures kept on their shelves can enable them to be proactive to include those requirements at the initial planning stage of the project rather than reactive to mishaps. Safety starts with you, and you need to be part of the team by seeing that safety requirements are observed in completing the job as scheduled, with quality and safety.

As the author of this book, I would like to encourage the reader, regardless of your position in the company, to go over the contents of this book, learn the safety fundamentals and best safety practices in the construction industry, widen your knowledge, and improve safety awareness of dynamic hazards so as to protect yourself and even your co-workers.

Thank you very much for purchasing this book, and wishing you all the best.

Introduction

This book is intended to help safety practitioners, project engineers, project managers, construction managers, and craftsmen strengthen their knowledge in safety as a construction company's priority in protecting the well-being of the people and the company's assets during the execution of projects.

The content addresses the safety fundamentals and best practices in the construction industry as outlined in each chapter—to wit:

1. Initial Project Planning
2. Overview of Occupational Health and Safety Management
3. Construction Management Assignment of Responsibilities
4. Organizational Structure
5. Lessons Learned from Unsafe Acts and Conditions
6. Hierarchy of Hazards Control
7. How to Conduct Safety Meetings Effectively
8. Developing a Hazard Identification Plan
9. Developing a Method Statement
10. Steps in Developing Job Safety Analysis (JSA)

This offers a wide understanding for both safety practitioners and company site management of the occupational health and safety requirements and international safety standards applicable in the construction industry, as prescribed by the Occupational Safety and Health Administration (OSHA) for the construction industry.

In spite of all known safety standards in the construction industry, many firms still have difficulty in meeting these requirements. The result is non-compliance and jeopardy to the health and safety of workers. One of the greatest challenges safety practitioners face is getting front-line supervisors and workers to fully understand the value of reporting unsafe acts and conditions, near misses, and incidents, with a corresponding negative impact on their performance in safety. As a result, the numbers of unreported incidents and the likelihood of injury to workers will most likely increase.

With my vast knowledge and experiences acquired from Saudi Aramco projects, the largest oil and gas plants in the world, and extensive training seminars from international safety associations, will help readers improve their safety awareness, regardless of their position in the company. The information provided in this book is intended to promote the understanding that any health and safety program must have the well-being of workers as its primary focus. The most important resource in any company is its people.

The regulations and safety standards referred to in this book are mainly the Occupational Safety and Health Administration (OSHA) standards. These standards serve to support company managers, project engineers, safety professionals, site personnel, and contractor or subcontractor employees.

The guidelines and procedures outlined in this book are the minimum requirements for occupational health and safety compliance and serve as best practices in the highly competitive global construction industry. As the author, I will accompany you to the real world of the construction field.

This book is intended as a primer, and I urge you to regularly increase your knowledge by engaging in continuing education.

Chapter 1

Initial Project Planning

This chapter will guide project managers, project engineers, safety personnel, and site supervisors regarding safety administrative requirements in preparation to tender their bids for construction projects through necessary phases, including the project-execution stage. It provides recommendations to follow the initial requirement at the planning stage of the project, in order to increase the likelihood of success in completing projects safely.

Chapter Objectives

As soon as you complete reading this chapter, you should be able to do the following:

- Describe the administrative framework of safety requirements to prevent injury, illness, and death of the workers.
- Identify the preliminary safety requirements to be submitted to the client's contract department for review and approval.
- Identify the standard operating procedures of submitting the contractor occupational health and safety program.
- Describe the client's internal safety requirements by the contract department.
- Recognize employee involvement at the early stage of the project to help develop the safety program.

Dr. Pedro P. Marfa, PhD, MBA, BSCE, OSH Consultant

Administrative Framework

Many workers are killed or injured every year as a result of workplace hazards. In human terms the sufferings are enormous, and the economic costs of the failure to ensure occupational health and safety are the top priority of a construction company to keep from undermining national aspirations for sustainable economic and social development. Improving occupational health and safety is in the best interests of all companies, employers, and workers. The measures to make improvements should be discussed and agreed upon. Successful health and safety practice relies on cooperation and goodwill as well as the safety protection of the workers concerned. It is hoped that this book will provide a basis for action to reduce the numbers of accidents in the construction industry.

The occupational health and safety regulations mentioned in this book are mainly related to the construction industry involved in government projects, the private sector, and oil and gas-producing facilities. My own expertise is in the fundamental safety requirements and best safety practices in the oil and gas construction industry. Knowledge of the regulations could help a professional in securing the offer of a better position in petrochemical plants as compared to jobs outside the oil and gas business. Likewise, it would help any construction company's effort to develop an occupational health and safety program required for pre-qualifications review prior to tendering bid proposals for major construction projects, as well as minor construction and maintenance work by oil and gas facility owners. In fact, a construction company with a good record in safety has an advantage over competitors in being awarded a project in a major oil and gas producing plant that could have a good project package.

Preliminary Safety Requirements

As soon as all the required documents are submitted to the contracts departments, whether in government projects, private firms, or oil-and-gas producing plants, the committee is obliged to instigate the review and acceptance process for project bidding and indicate the construction company categories. The contracts department sends the project bidding invitation to qualified construction companies to attend a job-explanation meeting for a certain project. The construction company representatives and project estimators normally attend this meeting, hosted by contracts representatives with the collaboration of the client project engineers and quality and safety representatives. Administrative contract requirements are presented in this forum, including a brief of the project's scope, terms and conditions, contractor responsibilities, and expected duration. The client project engineers explain the project scope in detail, engineering designs, procurement and delivery requirements, estimated project completion, partial mechanical completion of the project, and claims for billings and change-order regulations. In relation to the project requirements, the quality representative discusses the quality aspects and contractor's obligations. The client safety officer explains the occupational safety and health requirements of the projects and gives citations to be followed so as to comply with the current safety regulations, as well as the international safety standards applicable to the project scope. The client safety officer reminds construction companies to read and go over the contract-agreement clauses specific to safety requirements, so the contractor includes any costs pertaining to safety in the bid price.

Each client representative must deliver the required information to respective bidders. The contract representative and client project engineers will give an opportunity to construction company representatives to raise any questions related to the project scope for clarifications. It is noted in the forum that all questions, queries, and clarifications surfacing after the meeting shall be addressed to the contract representatives through communication means provided by the contracts department representative to receive the queries and answer all bidders in a specific time.

Right after the meeting and discussion, the client's project engineer has to schedule a site-visit walk-through with all the potential bidders to ascertain that points discussed during the explanation meeting will be physically evaluated at the jobsite and considered in their bid price. It also helps the construction company representatives assess the site conditions. These governing standard operating procedures are applicable to any government or private sector projects and oil and gas producing facilities where health and safety of the workers is a major concern. Since safety is an integral part of a company's success in completing projects, construction companies should take it seriously and equal in fulfilling a commitment to progress completion. Company representatives shall introduce best safety practices prior to beginning any of the construction activities.

Standard Operating Procedures

Construction companies around the globe aspiring to have a project in a challenging work environment with major construction in government, the private sector, and oil and gas producing facilities have to prepare and develop an occupational health and safety program (OHSP) based on the project scope. Apart from a safety program, the construction company shall also develop a detailed execution plan for the project as a baseline reference for the construction work, whether in the oil-and-gas-producing plants, private construction, or government infrastructure projects. Normally, the execution plan shall be prepared by project engineers in collaboration with project managers. The occupational health and safety program shall be developed by the health and safety manager, with the cooperation of safety officers and safety experts, to make sure that all applicable safety requirements from the client's internal safety procedures are included in the program; internationally accepted safety standards and local regulations shall also be incorporated and addressed in the program.

These documents shall be submitted to the facility owner contract department during the construction company pre-qualification reviews. Once the construction company meets the requirements and a bid is accepted, the company will be provided with the project scope of work by the facility owner, administered by the contracts department. In addition to the scope of work, the contract agreement also outlines the safety requirements for the projects. The construction company is mandated to comply with these requirements throughout the completion of the project. Specific health and safety requirements from government agencies, private sector firms, and oil and gas-producing facilities should be implemented as a priority to prevent employee occupational illnesses, injuries, or death.

Although the applicable project safety requirements have been provided to the construction company, there may be cases in which the company's contracts department does not consider the major cost of health and safety program requirements during the bid process. These unforeseen expenses may hinder the effectiveness and efficiency of the construction company to perform at client-expected levels upon commencement of the project. The most common failures are these:

- Non-compliance of health and welfare facilities for the workers
- Inadequate personal protective equipment
- Inadequate heavy equipment to be used in the project
- Lack of scaffolding materials
- Inadequate number of safety personnel assigned in the project
- Lack of competent persons assigned to the project
- Inadequate number of skilled workers
- Untrained workers assigned on the job

Failure to provide adequate resources hinders the company's ability to perform safely and damages the company's reputation, especially when the project's progress begins to decline and the workers begin to perform their assigned tasks unsafely in order to make up for delays. Best safety practices to prevent these conditions are addressed in the following chapter of this book.

Clients' Internal Safety Requirements

Contract clauses, such as administrative, quality, safety, and environmental requirements, are the minimum safety guidance provided to the bidders. The minimum guidance should be carefully addressed in the bid in order to meet client expectations for safe project execution. Failure to comply with these requirements may result in the client, or facility owner or operator, stopping work or cancelling the contract, depending on legal requirements set forth in the agreement between the construction company and the client. Any monetary loss due to delays is solely the responsibility of the construction company or the company's contractors or subcontractors, depending on the contract. Familiarity with client's minimum safety requirements is vital for the construction company to identify the requirements for the project during the bidding stage to avoid surprises from the construction company during project construction or as soon as the standard regulations in safety are enforced at the worksite by the client's safety officer. It serves as a reference to follow in performing any construction activities to protect the workers from being exposed to any potential hazards that may develop at the workplace.

Government agencies, private firms, and facility owners of the projects have their own safety personnel who will continuously monitor the construction activities to make sure that the construction company is implementing the minimum safety requirements and that the established rules and regulations are properly applied. Most construction company weaknesses rely on the assumption that they have understood the requirements rather than carefully reading the contract clauses under the health and safety regulations section. They focus on contract sections where detailed work description forms the basis of the project cost. The construction company should understand the requirement of every activity and corresponding safety regulations that each activity needs to address. If the contractor fails to perform proper understanding, the results can be construction delays, leading to cut corners on their activities as well as increases in incident rates, injuries, and occupational illnesses.

Employees' Involvement

Early involvement of every discipline in the project benefits both the company and the client carrying out the job safely and meeting client expectations. Ensuring the well-being of employees as well as protecting the company's assets and resources are very important goals to achieve. Safety should be given equal importance with quality and production or progress. Emphasizing progress at the expense of non-compliance with the safety rules and regulations on the project could lead to injury, property damage, and loss of production. A company's reputation can be damaged due to unsatisfactory performance, hindering the company's future bidding.

Construction companies should employ only competent personnel within each project discipline for the project. It is crucial to see that skilled engineers, safety officers, and workers will oversee and perform the construction activities and implement the company's occupational health and safety program, as previously reviewed and accepted by the clients. It is important to understand the scope of the project and plan to hire personnel in order to ensure adequate staffing to support the project. Each employee is responsible to adhere to safety guidance if the project is to be completed safety, within budget, and within schedule.

Because safety of workers and protection of company assets are critical to safe performance at the construction site, the safety officer must be actively involved in accident and injury prevention, and a culture of safety must be established from the earliest phases of the project through and including completion. Input and recommendations by the safety officer are key aspects of the health and safety program. The safety officer is a safety professional who is well versed in international health and safety policy, regulations, standards applicable for the project, and guidance and, as such, should be looked upon as a key component of the project management staff. Apart from being oriented with international and local safety regulatory guidance and standards, the safety officer also is well oriented with the company occupational health and safety program. Leadership and accountability of project managers shall be fully committed to health and safety program of the project in order to be successful. See chapter 3 for details of employees' responsibilities.

Chapter 2

Overview of Occupational Health and Safety Management

This chapter is discuss the minimum recommended safety requirements, standards and procedures of developing occupational health and safety requirement of the project. It provides a guidelines to support Construction Company to provide resources, competent persons, safety orientation, and safety meetings, reporting of incidents, and proper training of employees for project execution.

Chapter Objectives

After reading this chapter, you should be able to do the following:

- Describe how the general principles of occupational health and safety are integral to the construction industry.
- Recognize the occupational health and safety standards to be implemented in construction sites with the support of site management.
- Identify the minimum safety management system as guidelines in developing a construction occupational health and safety program.
- Identify the qualification requirements for safety personnel and safety trainings.
- Discuss the safety orientation requirements for new and transferred employees on the construction project.
- Recognize and be familiar with the proper reporting preliminary incidents report.

- Recognize the importance of safety meetings at the construction site to improve the safety awareness of the workers.
- Identify the importance of safety inspections at the construction sites.
- Recognize the advantage of good housekeeping at the construction sites.
- Identify the safety requirements in the contractor laydown yard for material and temporary site office.

General Principles

The promotion of occupational safety and health is part of an overall improvement in working conditions. It represents an important strategy, not only to ensure the well-being of workers but also to contribute positively to productivity. Healthy workers are more likely to have a higher work motivation, enjoy greater job satisfaction, and contribute to better quality outcomes in the construction industry in service to government agencies, private firms, and gas and oil producing facilities. This can enhance the overall quality of life of individuals and society. The safety and health of workers are essential to quality and production. Changing of work behaviour in the construction industry is of utmost importance for safety and production.

In order to ensure that satisfactory and affirmative results are achieved in the occupational safety and health requirements, every employee should aim to prevent injury, incidents, and illnesses arising at the workplace. By striving to mitigate the causes of hazards inherent in the working environment, this program will help reduce the costs associated with work-related injury and illnesses, thereby contributing to the improvement of working conditions and productivity. The implementation of this program will reaffirm the employer's commitment to a safe working environment and enable them to comply with the company safety program, which promotes employees' morale through effective action. Consistent implementation of the occupational safety and health program in all parts of the company can only be achieved through sustained and continuous efforts. It is the employer's and employees' team effort that should be exerted to have a safe and healthy working environment.

Given the complexity of construction projects for government, the private sector, and gas and oil producing facilities, safety infractions are the causes of occupational hazards and work-related illnesses. Practical safety measures may vary depending on the degree of employees' safety awareness and the resources available. It is possible, however, to give a broad outline of the essential components of occupational safety and health programs for construction projects.

In general, it should address in detail each employee's role and responsibilities in compliance with the company safety program aligned with the international safety standard regulations and client's safety program. Appropriate safety standard regulations applicable on each project scope, together with adequate means of enforcement, are the key factors to ensure that best safety practices are implemented to protect workers adequately.

Implementation of Occupational Health and Safety Standards

Occupational safety and health standards implementation should be supported by the company's upper management, including regular site safety inspections, management walk-throughs, and audits to make sure that regulations are enforced. The observations during the safety inspection or management walk-throughs should be assessed to support workers' health and safety compliance. The enforcement of the latter could be an integral part of the safety program of prevention, protection, and promotion of the company's personnel at site. Employers and line managers have to fulfil these objectives by adopting appropriate techniques, and the efficacy of safety measures ultimately rests on their application to the workers. It is imperative for company managers, site supervisors, and workers to be consulted as to whether the company supports them by providing the required resources to perform the job safely and complete the job within the allocated time frame.

Safety is everyone's responsibility; thus, all employees, regardless of their discipline, should collaborate in promoting a safe and healthy working environment. Each employee should understand that the success of the company and prevention of injury, illness, and death depend on their own action. Thus, employees should not rely on safety personnel reminding them to comply the requirements; instead, they themselves must take proactive action to follow the rules for their own protection.

It is highly unlikely that employees and site supervisors on the construction site come from different parts of the world, and there may be language barriers. Accordingly, the level of understanding and education of employees may vary. The safety culture and working environment in different countries are not the same. This requires a wide understanding for effective management in acquainting them with the present working environment. For instance, some workers may not know how to use power tools for carpentry works or how to use a fall-arrest system when working in elevated areas.

Many safety requirements, international standard codes, and local company standards could apply to the construction where the employees are working, and they need to be familiar with requirements, especially site-specific ones, prior to beginning their work on a construction site. An effective way to help both new and transferred workers to identify hazards is through a detailed safety orientation class provided by the company's safety officer. Aside from the orientation given by the company, the client's representative shall provide a safety orientation program from their facility. Most oil and gas producing plants or related facilities require all construction company employees to attend a mandatory safety orientation class.

Administrative Safety Management

Construction company administrative safety guidelines shall outline the minimum requirements to be implemented during the construction activities, regardless of the location of the project. Each facility has its own safety program in compliance with international standards and aligned to the facility owner's safety program. A developed occupational health and safety program has to be specific to the scope of work and meet the company owner's requirements. To ensure that the minimum safety standards and regulations are fully implemented at the workplace in accordance with the client's requirements, the construction company shall assign a highly qualified safety officer to oversee the company's safety program at the project sites.

Construction companies have different strategies of developing Occupational Health and Safety Programs, depending on the nature of the project and client's corporate safety program. It may vary sometimes due to the location of the project in compliance with the government regulations. As for the initial planning stage of developing the safety program, below are the recommended guidelines to address in the occupational health and safety program.

- Scope of work
- Company's safety policy
- Organizational chart with safety reporting relationships (refer to chapter 4, Organizational Structure, for guidance)
- Names and qualifications of safety staff
- Assignment of responsibilities by discipline
- Training needs analysis
- Written safety training program
- Hazard identification plan

Qualifications of Safety Personnel and Training Certification

Construction companies shall make sure that safety personnel are competent to supervise the implementation of the occupational safety and health program. Preferred candidates for safety personnel based on previous work fall into the following categories:

- Safety managers, safety superintendents, and safety advisers should have at least eight years of experience in the construction industry, be fluent in English, and communicate well in writing. They should be graduates of engineering

courses accredited by international safety organizations and have the ability to develop safety procedures for the company.
- Safety supervisors should have at least six years of experience in the construction industry, be fluent in English, and communicate well in writing. They should be graduates of four year courses and have accreditation from international safety organizations and they should have the ambition to implement a company safety and health program.
- Safety officers should have at least three years of experience in the construction industry, be fluent in English, and communicate well through writing. They should have at least college level training and be willing to enforce company rules and regulations.

They should have the required training and certifications related to tasks provided by the company. Implementation of a safety program at the construction sites is not easy, as people are often reluctant to comply with company rules and safety policies. It requires understanding of both the safety regulations within the company program and international standards applicable for construction.

There are two classifications of safety officer that we can find at the construction sites. (This may help you assess which classification you may belong to.) One of these qualifications of safety officers are those who can perform their job with continuous supervision from client safety representatives and have less initiative to implement the procedures and best safety practices without being told by their immediate supervisor or client representative. As they go along in the project, it becomes a burden to the company, which can lead to a non-compliance report from the client's safety representatives and even result in a work stoppage due to unsafe acts and conditions. Higher management may not have immediately noticed the shortcomings at the construction site, but the safety representative from the client understood the qualifications and competency of the safety officers assigned in the project.

Some companies assign safety officers to the construction site just to satisfy the client's safety requirements. These safety officers may not be able to perform their duties and responsibilities as expected by the client due to their limited knowledge and understanding of fundamental construction safety requirements applicable in the project. Instead of implementing the rules and regulations and giving recommendations to correct unsafe acts and conditions in the construction site for safety compliance, they tend to do a job which is not their responsibility. This may be fixing barricades, collecting or delivering drinking water igloos to the jobsite, and even using the site supervisor to deliver construction materials from one site to another. In most cases, some of them are doing their best to identify the hardware aspects of safety, such as seeing workers without personal protective equipment (e.g., safety glasses, helmets, or safety shoes) applicable for their job.

Safety is everybody's responsibility, and it is a common objective shared by all employees in the company. Everybody is expected to comply with these requirements for their own protection, to prevent injury or occupational illnesses acquired from construction hazards. The recommended approach is supporting the implementation of company occupational safety and health program through a regular safety toolbox meeting and addressing the potential hazards and their mitigation prior to starting the workday. Specific job skills training is required as continuing education of the employees.

The second classification of safety officers at the construction site are those who are knowledgeable and familiar with the client's safety program. They understand the best safety practices in construction and are well oriented to the international safety standards applicable to construction projects. A competent safety officer can do a great deal to support the implementation of a company's occupational health and safety program. Apart from being oriented to the company's safety requirements, they can help the company develop a risk assessment program, hazard identification plan, and specific procedures for the critical work activities in the field.

Upon knowing these two different classifications of the safety officer, you can assess yourself as to the category in which you belong. It is essential to the construction company that the safety officers assigned to oversee the implementation of the company's safety and health program are conversant in safety standards, rules, and regulations of both the hardware aspect of safety and the standard safety regulations to be implemented in the project sites.

Employee Safety Orientation

Construction companies shall provide a safety orientation class to all workers prior to mobilizing them at the jobsite. The safety orientation should be given to the newly hired employees as well as to those newly transferred from other project site locations. It is required because the potential hazards from previous project locations may not the same as those pertinent to new project assignments. The orientation topics to be discussed should include but are not limited to the usage of personal protective equipment (PPE), housekeeping, emergency response plan procedures, internal safety procedures of the company, job skills training, and the requirement of permit to work system from client's requirements. Each facility, whether in a construction area or an oil and gas producing facility, has its own unique procedures with which every employee is expected to comply.

Contractor companies working in projects for oil and gas producing facilities should emphasize to workers the restrictions of using mobile phones, driving safety, emergency evacuation drills, and smoking. Any owner-established designated evacuation area and other restrictions and procedures must be followed.

Preliminary Incident Report

As much as possible, potential hazards in the construction site must be controlled to prevent incidents. However, if the conditions are beyond control and something happens on or off the job, proper reporting procedures must be followed. Timely reporting of incidents is required to make sure on-time response to the exact location of incident occurs as appropriate. A preliminary incident reporting guideline should be included in a construction company's occupational safety and health program, and every employee in the company should be oriented and familiar with this procedure.

Reporting of incidents, regardless of whether they are major or minor, shall be directed to company site management and to the facility owner for their records. It should be emphasized during the safety orientation class that reporting of incidents to higher management will not adversely affect their performance rating; instead, it gives a positive indication that the construction site is free from potential hazards to the workers.

Employees involved with an incident, whether on or off the job, report the incident verbally to his immediate supervisor if he is able, and it should be done immediately. If for any reason the injured person is not be able to do so, then the immediate supervisor should notify company management as soon as the injured employee is stable, after he has received first aid treatment. As soon as the verbal report has been completed, the employee shall prepare the written preliminary report with the assistance of his supervisor and submit it to the division head within twenty-four hours after the incident. Then the company investigation committee shall conduct the investigation to find the root cause of the incident. A detailed investigation report shall be submitted to the division head, client safety representative, and department manager within thirty-six hours after learning about the incident.

Once the incident report is reviewed and accepted by the department manager, an action prevention memorandum must be developed and shared with all division personnel as a lesson learned. A concise action bulletin should address recommendations for improvement and steps to prevent recurrence of the incident.

Safety Meetings

A two-way method of communication is a crucial practice to address observations at the construction site. Unsafe acts and conditions and near-misses should be brought up as topics during the discussion in the safety meeting to raise workers' level of safety awareness and promote a safe working environment and worker protection. A safety meeting should take place at least once a week for thirty minutes. This meeting should be spearheaded by the construction site supervisor with the assistance of the company's safety officer. The safety meeting shall be attended by all company employees so that all are fully aware of the safety requirements in the construction site.

Previous safety observations and safety requirements for upcoming construction activities shall be discussed in the meeting. Employees who have safety concerns at the jobsite can bring them up during the meeting for discussion and clarifications.

Safety Inspections

A regular site inspection at the construction area is a good tool to identify and correct potential hazards in the project site. Effective safety management should include upper management in the company whenever safety inspections will be conducted. This indicates to the company that safety is the topmost priority in the construction to protect the well-being of workers and prevent damage to company assets. Safety inspections can be done daily, weekly, and/or monthly, depending on the site conditions.

Daily safety inspection shall be conducted by the foremen and safety officer assigned to the project. Any observations made during the site walk-through inspections shall be recorded and forwarded to the construction division for immediate corrective actions. Observations that cannot be corrected immediately must be entered in the inspection log sheet and followed up in following days for the updates. A proactive approach in safety to correct the deficiencies noted during the walk-through helps keep the jobsite safe for the workers.

Weekly inspections shall be conducted by a team composed of foremen, site supervisor, project engineer, and safety officer assigned to the specific site. Safety observations during the walk-through shall be recorded and submitted to the division head of the company. The resources required for the corrective action shall be coordinated by upper management for immediate action. A regular weekly inspection helps to promote occupational health and safety protection of the workers.

Monthly inspection walk-throughs shall be conducted by a team composed of construction manager, project engineer, site supervisor, and safety manager of the company. Participation of higher management in safety inspection provides positive outlook for the company and shows that the management is committed to support the implementation of the program. Safety observations that were not corrected immediately during the walk-through shall be entered in the inspection log sheet for record and appropriate action. A follow-up inspection is required for corrective action and closure.

All safety observations of the projects shall be consolidated in a monthly report for trending analysis for improvements. Additional safety inspections may be required as directed by the management.

Housekeeping

Housekeeping is everyone's responsibility at the construction site, and it is of prime importance as work progresses. The workers may misunderstand the benefits and importance of housekeeping in day-to-day activities in the construction site. Poor housekeeping at the workplace increases the chances of exposing workers to potential hazards associated with construction activity. The most common safety hazards encounters at the jobsite are trips, slips, and falls. Job-related injury may occur when workers step on nails protruding from lumber.

Some workers refuse to participate in housekeeping tasks because they are not labourers. They see cleaning up as a labourer's job only. Lack of understanding is what drives this attitude, and all employees must understand that housekeeping is a safety-related activity. Accidents can easily happen when the housekeeping in the area is uncontrolled.

Just to provide a simple example. There are twenty four hours in one day, and your time at work in one shift is eight hours. Sometimes the working time will be extended up to ten hours, including overtime, depending on the urgency of construction activities. So after a ten-hour workday is subtracted from twenty-four hours, fourteen hours remain. Set aside at least eight hours for sleep. The remaining six hours are free time to move around in your camp accommodation. Out of six hours, three hours are using for house chores while the other three hours are utilized to take a break in the recreation room or watching television.

To summarize, employees are doing some routine jobs at home for three hours each day, with less exposure to any hazards. So if the camp accommodation is organized and free from safety risks, then there is every reason that housekeeping at the jobsite should be conscientious enough to avoid slips, trips, and falls. It's a big challenge to all construction workers to maintain a good housekeeping at the jobsite.

Contractor Laydown and Fabrication Yard

Initial planning shall be done at an early stage of the project to identify the location of all construction materials for the project. A ground plan of the laydown yard should normally convey the following information in the drawing:

- Identify the location of combustible and flammable materials
- Indicate the separation distance between the combustible and flammable liquids
- Identify the location of temporary support facilities or bunkhouses during the construction
- Indicate the separation distance of buildings, either permanent structures or temporary portable buildings, as per international building code (IBC)
- Identify the location of the fabrication shop and indicate the distance from temporary support facilities
- Identify the location of firefighting equipment such as fire hydrant, monitor, fire hose reel, and fire extinguishers
- Indicate the staging area for demolished and de-nailing areas
- Identify the smoking area in the laydown yard

The plan drawing, reflecting this information, shall be submitted to the safety department in the construction company for review prior to send the documents to client's safety representative for further review and acceptance. As soon as the client safety representative accepts the proposed plan for the laydown yard, it shall be implemented prior to receiving the construction materials and beginning the construction.

When the construction activity begins, the company shall designate a full time supervisor to oversee the laydown yard material storage to avoid any disorder or bad housekeeping.

Many workers on the construction site may not know the benefits of good site housekeeping. Maintaining good housekeeping anywhere on the worksite and laydown yard can prevent incidents leading to injury, keeping workers safe and preventing bodily harm.

Apart from this, it also helps speed up the handling and delivery of construction materials.

Garbage bins with plastic liner bags should be provided to make garbage collections easier. Each garbage bin should be provided with a lid and must be closed at all times to prevent clutter and keep the insects and rodent away from the bins. Regularly scheduled collection of the garbage bin is essential to prevent accumulations that may pose fire and health hazards.

Chapter 3

Construction Management Assignment of Responsibilities

This chapter outlines management assignment responsibilities to make sure that all employees in the construction company know their roles in supporting the company goals and objectives in protecting employees' welfare and company assets. It also defines the company policy, guidelines, and responsibilities of the employer and all employees in the company in compliance with Occupational Safety and Health Administration and other applicable standards in construction industry.

Chapter Objectives

After reading this chapter, you should be able to identify and understand your responsibilities as an employee in the company:

- Discuss the management commitment in the compliance of occupational health and safety program.
- Recognize the company health and safety policy applicable in the construction industry.
- Recognize and understand the employer responsibilities to the workers.
- Identify the project manager responsibilities in the company occupational health and safety program.
- Identify the safety manager/safety supervisor responsibilities as outlined in the occupational health and safety program.

- Describe the project engineer, site safety supervisor, and foreman responsibilities in the construction site as outlined in occupational health and safety requirements.
- Identify the responsibilities of the workers in the company.
- Discuss and describe the employee workers cooperation of the implementation of occupational health and safety program.

Management Commitment

While top management has the ultimate responsibility for the health and safety program in the company, authority should be delegated to all management levels to ensure safe construction practices. Supervisors are obviously the key persons committed to enforce the implementation of such a program, because they are in constant contact with the employees. As safety officers, they act in a staff capacity to help administer safety policy, to provide technical information, to help with training, and to supply program materials. Full commitment on the part of construction management to making health and safety a priority is essential to a successful occupational health and safety program in the workplace. It is only when management play a positive role that workers view such programs as a worthwhile and sustainable exercise. The management has the influence, power, and resources to take initiative and to set the pattern for a safe and healthy working environment.

Management commitment to occupational safety and health may be demonstrated in various ways, such as:

- Allocating sufficient resources for proper functioning of the occupational safety and health program.

- Establishing organizational structure to support managers and employees in their occupational health and safety duties and responsibilities.
- Designating a senior management representative to be responsible for overseeing the proper functioning of occupational health and safety management.

The process of organizing and running an occupational safety and health program requires substantial capital investment. To manage safety and health efficiently, the financial resources must be allocated within the business units as part of the overall running costs. Management must understand the value that the corporate leaders place on providing a safe place of work for employees. There should be incentives for managers to ensure that resources are deployed for all aspects of health and safety. The challenge is to institutionalize health and safety within the planning process.

Health and Safety Policy

Since occupational accidents are work-related injuries of individuals at the workplace, prevention and control measures within the company should be planned and initiated jointly by the employer, managers, and workers concerned.

Well-defined policy represents the foundation upon which occupational health and safety goals and objectives, performance measures, and other system components are developed. It should be concise, easily understood, approved by the highest level of management, and known to all employees in the organization. The policy should be in written form and should cover the organizational arrangement to ensure health and safety. In particular, the policy should:

- Allocate the various responsibilities for occupational health and safety within the company or organization.
- Bring policy information to the notice of every worker, supervisor, and manager.
- Determine how occupational health services are to be organized.
- Specify measures to be taken for the surveillance of the working environment and workers' health.

The policy may be expressed in an organizational mission and vision statement, as a document that reflects the company's health values. It should allocate the various responsibilities of each department head in the company, through the line management, project engineers, site supervisors, and the safety and health coordinator, who will be the prime mover in the process of translating policy objectives into reality with the company.

The company policy must be printed in a language or medium readily understood by the workers. Where illiteracy levels are high, clear non-verbal forms of communication must be used. The policy statement should be clearly formulated and designed to fit the particular organization for which it is intended. It should be circulated to all so that every employee has the opportunity to become familiar with it. The policy should also be prominently displayed throughout the workplace to act as a constant reminder to all. In particular, it should be posted in all management offices to remind managers of their obligations in this important aspect of company operations. In addition, appropriate measures should be taken by the competent authority to provide guidance to employers and workers so as to help them comply with their responsibility. To ensure that the workers will accept company's safety and health objectives, the employer should establish the policy through a process of information exchange and discussion with them.

Reviewing a policy statement is also necessary to keep it alive and aligned to any new requirements. A policy may need to be revised in the light of new experience or because of new hazards. Revision may also be necessary if the nature of work to be carried out changes or if new hazards are introduced into the workplace. It may also be necessary if new regulations, a new code of practice, or amended official guidelines are issued that are relevant to the activities of the company.

Employer's Responsibilities

The company policy should reflect the responsibility of employers to provide a safe and healthy working environment. The measures that need to be taken will vary depending on project locations and activity and the type of work performed in general. OSHA Standards for the Construction Industry citing employer and Rule 1040, Occupational Safety and Health Standards, Department of Labour and Employment, Philippines, are citing employer and employee responsibility. It should include but not be limited to the following:

- Provide and maintain workplaces, machinery, and equipment, and use work methods, which are safe and without risk to health as is reasonably practicable. As per hierarchy of controls (Administrative Controls, Engineering Control, and Personal Protective Equipment [PPE]).

- Provide the necessary instructions and training to managers and staff, taking account of the functions and capacities of different categories of workers.
- Provide adequate supervision of work and of the application and use of occupational health and safety measures.
- Institute organizational arrangements regarding occupational safety and health, adapted to the size of the undertaking and the nature of its activities.
- Provide adequate personal protective clothing and equipment without cost to the worker, when hazards cannot be otherwise prevented or controlled.
- Ensure that work organization, particular with respect to hours of work and rest breaks, does not adversely affect occupational safety and health.
- Take all reasonable and practicable measures to eliminate excessive physical and mental fatigue.
- Provide, where necessary, for measures to deal with emergencies and accidents, including adequate first aid arrangements.
- Undertake studies and research or otherwise keep abreast of the scientific and technical knowledge necessary to comply with the obligations.
- Cooperate with the employers in improving occupational health and safety practices.

Project Manager Responsibilities

The project manager is accountable to top level management for taking action on the implementation of occupational health and safety program. He is responsible for providing the required resources for the implementation of the program and ensuring the safety performance and accountability of subordinates. He is also the first line of reporting for the construction manager, project engineer, and site supervisors.

The project manager is also responsible for reviewing and acting on safety reports from line supervision and for communicating to subordinates regarding corrective action and worker protection. His responsibilities include but are not limited to the following:

- Ensure that personnel are in compliance with the company occupational health and safety program.
- Initiate the company policy for the control of injury, damage, and fires.
- Administer the policy himself, and appoint a senior member of staff to do so.
- Know the client safety requirements, and ensure that the workers will observe them.
- Ensure that all supervisors are qualified and receive adequate training.
- Coordinate safety activities between the client and the construction company.
- Observe the process of investigation and proper reporting procedures.
- Initiate analysis and determine the root causes of any accident.
- Ensure that a structured hazard identification plan shall be developed and implemented.
- Review and respond to safety reports.
- Maintain effective communication on safety matters to all line supervisors, and include safety as part of each project meeting.

Safety Manager Responsibilities

- Advise management on ways to prevent injury to personnel and damage to facility and equipment.
- Advise ways to improve existing work methods.
- Advise the legal and contractual requirements between the client and contractor safety requirements.
- Advise the potential hazards on site before the work starts.
- Carry out the site surveys, and observe that safe methods of construction are followed.
- Assist with training employees at all levels.
- Keep up the record up to date, and circulate the information to all workers.
- Liaise with client safety representative to conduct safety audit.
- Attend the weekly progress meeting where safety is the agenda of discussion.

Project Engineer Responsibilities

- Understand the company policy and responsibility to enforce the workers' compliance.
- Know the client's safety requirements and expectations, and ensure that they are observed.
- Provide written instructions to establish work methods, explain the sequence of operations, and outline potential hazards at each stage.
- Verify the work methods and precautions with supervision before work starts.

- Develop a safety awareness campaign, promote safety meetings and presentations, and implement safety training.
- Set a personal example on site by wearing appropriate protective clothing and equipment at all times.
- Ensure that all accidents are reported in accordance with client's requirements, and provide root cause analysis of each accident.
- Attend and participate in the weekly safety meetings, site safety inspections, and audits.
- Report any unsafe acts and conditions to higher management.
- Liaise with the government representative to conduct safety inspections and audits.
- Ensure that good housekeeping at the jobsite is being observed.
- Ensure that employees are using proper tools and equipment for the job and wear personal protective equipment (PPE) at all time at the construction site.

Site Supervisor/Foreman Responsibilities

- Organize the jobsite so that work is carried out safely.
- Understand client safety requirements, and enforce compliance.
- Give precise instructions on responsibilities for the correct work methods.
- Conduct housekeeping before the end of each workday.
- Conduct a daily inspection of all power and hand tools, and make sure that they are in good condition.

- Make sure that all the workers are familiar with the evacuation area.
- Ensure that adequate personal protective equipment is available at the construction site.
- Send the employee when necessary to attend safety training.
- Cooperate with the safety officer when conducting incident investigations.
- Participate in accident/incident investigations, and submit investigating report to the supervisor.
- Conduct and participate in toolbox meeting prior to starting the job.

Workers' Duties and Responsibilities

The cooperation of workers with the company is vital for prevention of occupational accidents and illnesses. The company policy should therefore encourage workers and subordinates to play their essential role in this regard and ensure that they are given adequate information on measures taken by the employer on all aspects of safety and health associated with their work. The company policy should outline the individual duty of workers to cooperate in implementation of occupational safety and health policy within the company. In particular, workers have a duty to:

- Take reasonable care of their personal safety as well as that of other persons who may be affected by their own acts.
- Comply with instructions outlined in the company safety program for their own safety and health protection and those of others, and follow all prescribed procedures.

- Use safety devices and protective equipment correctly.
- Report unsafe acts and conditions to their immediate supervisor, including any situation which they have a reason to believe could present a hazard and which they cannot themselves correct.
- Report any incident/accident or injury which arises while at work or in connection with their duties.

Workers' duties in hazard control have as their counterpart the recognition of certain basic rights, and these should also be reflected in the company policy. In particular, workers should be assured of the right to stop work in the face of potential danger and to refuse to carry out or continue on tasks that they reasonably believe presents an imminent and serious threat to their life or health. They should be protected from unforeseen consequences of their actions.

Access to safety and health information is a prime condition for positive contributions by workers in their respective work locations to control occupational hazard. The company policy should make sure that workers are able to obtain any necessary assistance in this regard from upper management, who have both the full authority and clear responsibility to be involved in anything that concerns the protection of the life and health of the workers.

Workers' Cooperation

Cooperation between management and workers within the company is an essential element of the organizational measures that need to be taken in order to prevent incidents, accidents, and illnesses at the workplace. Participation is a fundamental worker's right as well as a duty. Employees have various obligations with regard to providing a safe and healthy workplace, and in the course of performing their work, their cooperation with the employer in the field is crucial to successful safety and health management and a major contributing factor in preventing and reducing occupational illnesses and injuries.

The full cooperation of workers in any occupational safety and health program benefits themselves. Such cooperation will not only ensure the effectiveness of health and safety measures but also make it possible to sustain an acceptable level of health and safety at a reasonable cost.

Workers' continuous involvement in improving the work process is vital, and it is only possible if everyone involved is properly trained. Training is an essential element in maintaining a healthy and safe workplace and has been an integral component of safety management for many years. Managers, supervisory staff, and workers all need to be trained. Workers should be given appropriate training in occupational safety and health requirements. It is up to management to give necessary instructions and training, taking into account the functions and capacities of different categories of workers.

The primary role of training in occupational safety and health is to promote action. It must therefore stimulate awareness, impart knowledge and help workers adapt safety awareness to their own roles. This training should not be treated in isolation but should feature as an integral part of job training and be incorporated into daily workplace activities. Management must ensure that all personnel who play a role or part in the construction process are trained in the technical skills that they need to do their work safely.

Chapter 4

Organizational Structure

The author recommends adoption of the organizational structure presented in this chapter. It has been used very effectively in many overseas projects. The construction company may adopt this organizational structure as a strategy to optimize teamwork. It also provides guidance to line management in the company to be implemented at the construction site.

Chapter Objectives

After you have read this chapter, you should be able to describe the following:

- Discuss the key principle of health, safety, and environment in the projects.
- Identify the appropriate chain of command in the organization of the line management.
- Describe the difference between centralization and decentralization management strategy in the construction company.

Key Principle of HSE

An effective organizational structure that I have used in previous projects is the chain of command extending from upper management to the workers on the construction site. In order to complete the process and clarify the expectations that everyone will comply with company rules and policies, the chain of command shall not solely focus on the safety professional for the implementation of policies as desired by the construction company. Such responsibility also applies to upper management personnel in the construction industry and in any business where personnel are exposed to a high-risk working environment. The communication flow from upper management to the workers is essential to the leader's accountability to oversee the implementation of the company safety program. Good organizational structures will promote communication and cooperation between workers and their immediate supervisor.

Occupational health and safety programs are part of an extensive, multidisciplinary field, invariably touching on concerns specific to various industries. Despite this variety of concerns and interests, certain basic principles can be identified and enforced. The functions and responsibilities of every employee in the company should be defined in the company's occupational health and safety program (see chapter 3 of this book, Management Assignment of Responsibilities, as a reference for implementation).

The employer shall designate a qualified project manager to oversee each department in the company. The assigned managers of every discipline, including the safety manager, are considered as the line management of the company and report directly to the chief executive officer (CEO) of the company or to the project manager as designated by the CEO. The level of authority of each manager is equal with other fields of discipline, and they will work in parallel as a team to implement the company occupational safety and health program. The safety and quality of the company are normally the focus of a separate department or division in the company organization, reporting directly to the CEO or project manager as directed by the CEO, to maintain independence from other managers.

Dr. Pedro P. Marfa, PhD, MBA, BSCE, OSH Consultant

Structured Safety Organizational Chart

Organizational structure is important for the company and must be developed prior to the start of any construction activities, whether government projects, private sector enterprises, or oil and gas producing facilities. Identifying the requirements in every discipline ahead of time will allow the company to prepare the appropriate manpower for the project who will effectively perform their functions in the organization. The organizational structure of the company shall indicate the management hierarchy of leadership and accountability to subordinates for the implementation of the company's safety policy and program. Since the employer's interest is to protect workers and company assets, the CEO shall designate a highly qualified manager in every discipline of the project to oversee the business operation. The recommended appropriate organizational structure is shown in Figure 1. Following this structure will help in implementing the safety program and effectively train the workers on their respective work and safety responsibilities.

Recognizing that the protection of the worker's life and health is a fundamental right, it follows that decent work implies a safe working environment. Furthermore, workers have a duty to take care of their own safety, as well as the safety of anyone at the worksite who might be affected by what they either do or fail to do. This structure will help enable them to know the required action to protect the people at the worksite and empower them to stop work in case of imminent danger to life and health. In order to take care of their own health and safety, workers need to understand occupational risk and dangers. They should therefore be properly informed of hazards and adequately trained to carry out their work safely. To maintain a safe and healthy working environment in the company, workers should cooperate with the employer to promote workplace safety and participate in implementing preventive measures. Because dynamic occupational hazards arise at the workplace, it is the employer's responsibility to ensure that the working environment is safe and healthy. This also means, however, that workers must cooperate to prevent incidents and protect themselves and other workers from potential hazards at the worksite.

Management visibly leading by example is essential to the company's success. They are to specify goals and set an example to ascertain that adequate resources, as well as qualified, trained, and skilled personnel, are available at the jobsite. Implementation of adequate health and safety compliance in construction and other industries requires management support.

Construction companies may adapt the organizational structure that have been implemented in my previous projects.

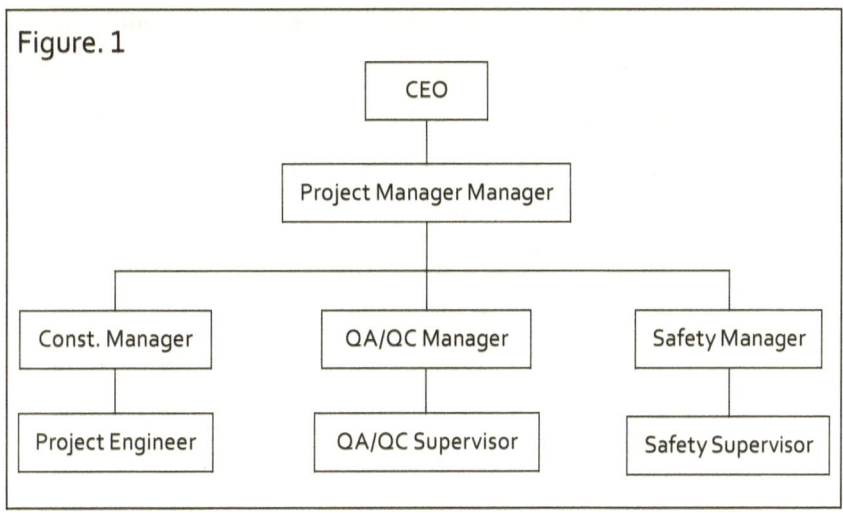

Either of two different cores of management functions in the construction site may apply.

Dr. Pedro P. Marfa, PhD, MBA, BSCE, OSH Consultant

Centralized Management

Each year construction projects around the world have increased, giving great opportunity to construction companies to offer their services in government projects, private sector projects, and oil and gas producing facilities. Many of these projects require numerous construction companies to build their facilities. For small scale contractors, in the beginning of the project, centralization management is the most effective way to run the business since employees have a direct communication from upper management level and among others in the construction workforce. This is a clear advantage of having a small number of personnel in the company. The communication of each worker is very efficient to take immediate action and provide the required resources to health and safety protection of the workers.

This advantage of centralized management extends to administering the implementation of occupational safety at the construction site. The process of supporting employees' needs will be efficient since upper management can learn immediately of the needed supports to workers in performing their duties and responsibilities at work. Another advantage of centralized management is that a small group of employees could be easily teamed up when developing, enforcing, and implementing the occupational health and safety program in the company.

However, the implementation of the program may not be as effective as expected in terms of compliance if line management – e.g., the project engineer or site supervisor – has a dual function at the construction site due to a limited number of employees to oversee the program. Sometimes this practice can confuse the site supervisor about which way shall be given more priority in the construction. Most of the time the site supervisor has to choose his priority in the construction between progress and safety protection of the workers from potential hazards at site, especially when the construction progress is far behind. The complexity of construction projects can be challenging and require care in handling the dynamic hazards associated with the work. Exposure of the workers to a hazardous environment with no remedial corrective action would lead to an increased number of employees involved with incidents.

Centralized management can have difficulty in undertaking bigger projects because of the company's limited resources and manpower. Most companies operated by centralized management are normally working on small scale projects or as subcontractors from bigger construction companies on major construction projects.

Decentralized Management

Decentralized management is a setup in which individual personnel have their own responsibility in each department or division to work for the company. As a company grows, decentralized management is the appropriate means of handling the construction progress, especially when handling multiple projects in a given location. The employees in every discipline have their individual responsibilities to develop procedures pertaining to their area of specialization in the construction field. The structure that has been developed was derived from the CEO perspective down to the project manager, construction managers, and quality and safety staffs within the company to support the construction activity. The direction from upper management shall communicate through the project managers, construction managers, project engineers, and subordinates down to the workers.

The advantage of decentralized management is the capacity to undertake more projects, up to the optimum strength of the company, as long as the resources to carry out the jobs safely at the construction sites are properly monitored and funded. The projects shall be handled by qualified personnel assigned to every department. Response from upper management is not as quick as from centralized management, but it would be beneficial within the organization of the company. The progress of the construction activities could not be the same as centralization management, but the efficient implementation of occupational health and safety is the top priority. Each construction company has its own specialization, while some other contractors are just working under the direct supervision from the prime contractors.

Chapter 5

Lessons Learned from Unsafe Acts and Conditions

In this chapter you will learn the basic examples of lessons learned about unsafe acts and unsafe conditions that you may encounter in the future projects. It ought to help you figure out how to improve your current strategies and resolve future problems effectively. It also gives more emphasis to handling the safety conditions by providing corrective action proactively without a cost.

Chapter Objectives

After reading this chapter, you should be able to carry out the following:

- Recognize potential hazards that may be encountered during the execution of the project.
- Identify unsafe acts and conditions at the construction site to prevent injury of the workers.
- Describe and discuss the recommendations of corrective actions to prevent recurrence.
- Implement a proactive approach to safety in any activity at the construction site.

Many construction contractors assume that their company is familiar with the safety standards and requirements of the general construction industry. Similarly, employees believe in handling projects in as they have in the past when it comes to implementing safety measures. It is of utmost importance for contractors to go over the company safety policy and regulations, not to overlook international standards applicable for the current project. This will help the contractor and key personnel to proactively prepare before bedding for projects. All lessons learned from previous projects should be kept in mind when developing a safety program; following safety recommendations and corrective actions from those lessons learned can improve performance and safety more successfully in future projects.

The maxim should be "Do it right in the beginning" in order to protect the welfare of the workers and their morale on and off-the-job. The well-known motto "Safety first" doesn't work in construction sites anymore because it gives a false signal to the worker by hiding unsafe acts and conditions at the construction site to protect the orchestrated motto. The welfare and protection of the workers are precious above all; thus, the employer shall provide a safe working environment at all level of activities. It is not just about giving it priority over progress. Maintaining the health and safety of the workers should be the most important tool to motivate the workers to improve the level of safety awareness at the construction sites. It will enhance workers' performance, making them more productive and eventually meeting project requirements and completion on time.

Another way of encouraging employees to improve safety awareness is to let them attend training for their individual craft. Such continuing education will improve their skills and affect their behaviour as self-motivated employees working harder with the highest level of safety awareness at work. These and other approaches work together to protect the well-being of workers and company assets.

This chapter will address common examples of hazards incurred at construction sites in which workers often overlook the importance of safety in their day-to-day activity. There are many unsafe acts and conditions at the construction site that you will observe when conducting site safety inspections. These unsafe behaviours committed by the workers are most likely to happen when the site supervision is not highly visible. Unsafe acts and conditions observed in construction sites must be reported to site management and immediately corrected. The corrective actions shall be discussed by all the employees in the toolbox meeting for their awareness.

The unsafe acts and conditions explained in this book include the following:

- mobile scaffolds
- no access ladder attached with free-standing scaffold
- working at heights
- crane and rigging operations
- portable ladders
- street ladder
- working inside modified man basket
- excavation
- power and hand tools
- working at heights without proper fall protection

Unsafe Acts/Conditions of Mobile Scaffolds

In most of the construction industry, the mobile scaffold is essential to enable workers to move easily from one place to another to perform their job. Any unsafe acts and conditions as shown in the picture could lead to an accident and may cause disability. Falling, tripping, and overturning a scaffold on a construction site is a major concern of the employer. Moving a mobile scaffold from one place to another while the worker is on the platform is considered an unsafe act, requiring corrective action. The employer shall provide training to the workers, emphasizing the potential hazards and consequences of unsafe practices. Employee's participation during the training is required for them to understand the danger of pushing a mobile scaffold from one place to another with workers on the platform.

Recommendations

The best safe practices in using a mobile scaffold shall be included when developing a procedures and safety program. It is beneficial to refer with the Occupational Safety and Health Administration (OSHA) CFR 1926, Subpart L safety requirements as guidelines to be followed by the employer during the execution of the job. It is the employer's responsibility to make sure that minimum safety requirements for use of a mobile scaffold shall be met prior to beginning any construction work, to address the following:

a. Mobile scaffolds shall be plumb, level and square with horizontal and vertical braces in both directions.
b. Horizontal bracing shall be installed at the base lift and top lift of all mobile scaffolds to prevent from racking.
c. Caster wheels shall be fitted with a positive wheel lock which cannot accidentally release, to prevent movement while the mobile scaffold is being used.
d. Caster wheels shall be securely connected to the post by using locking pins to prevent accidental release.
e. Caster wheels should be at least five to seven inches in diameter.
f. Caster wheels shall be locked at all times to prevent movement, except during movement of the mobile scaffold from one place to another.
g. The height of mobile scaffolds shall not exceed a maximum of 12.2 metres (forty feet).
h. Mobile scaffolds shall only be used and moved on surfaces sufficiently firm and level to ensure their stability.
i. A mobile scaffold shall be moved manually by pushing at the base with no workers on the platform.

Unsafe Conditions of Free-Standing Scaffolds

A scaffold platform is a means of access to perform a job safely on an elevated area. Abusing the usage of working platform apart from the workers is the most common unsafe practice at construction sites. Using the side section of the scaffold for access to a free-standing scaffold is an unsafe act. Loading the platform with construction materials such as concrete bricks, hollow blocks, cement bags, sandbags, or pipe spools, and for storing coiled electrical cables during cable pulling, can cause the scaffold platform to collapse. It is essential for construction contractors to assign a full-time competent person to oversee the scaffold erection and dismantling to continuously monitor the stability of a scaffold. Comprehensive training for personnel involved in elevated works is essential to familiarize them with the hazards associated with elevated construction activities.

Trips, slips, and falls are the most common adverse incidents occurred in the construction industry which construction contractors overlooked to prevent the workers from being harmed, which could cause them to be away for a significant time. Absenteeism of the workers from their normal duty would impeded the ability of construction contractors to complete the job in a specified time, risking the loss of a client's trust and any opportunity of being offered more projects by the client.

OSHA CFR 1926, Subpart L, has outlined the scaffold requirements regarding materials and methods of erection and dismantling, to ensure that scaffold erection shall meet the standard requirements that provide a safe working platform for workers.

Recommendations

The employer shall assign a full-time competent person to supervise construction activities which involve a tower or free standing scaffolds that will be used for block works and other activities and see that they meet the following scaffold criteria:

a. Scaffolds shall be designed by the structural engineer to make sure they can withstand the loads to be placed on the platform.
b. Scaffold components shall be capable of supporting, without failure, their own weight (dead load) and at least four times the maximum intended load (live load).
c. Scaffold shall have a specified load rating, corresponding to the maximum intended load as per the design calculations.
d. Scaffold components shall not be loaded in excess of a design rating as shown in the scaffold tags.
e. The employer shall designate a competent person to perform the design of any scaffold platform that will be used for storing construction materials.

Unsafe Acts of Working at Heights

Workers are placed in considerable danger when there is not adequate supervision at site to direct them to use the right working platforms or use full body harness and secure the lanyard to a suitable structure when working at heights. If workers carry out the assigned tasks on their own initiative, thinking the job can be done using unsafe equipment (or using it in an unsafe manner), most often the workers commit such unsafe acts due to lack of access to required resources. Adequate site supervision can also prevent shortcuts that risk the serious consequences of a fall. Training employees to identify the right equipment to perform the job is essential prior to assigning any work at the construction site.

Occupational Safety and Health Administration (OSHA) requires employers to provide a safe working environment when there is a potential for falls. Employees may assume that this is a routine job, similar to work they have done many times without mishap. Any incidents resulting from unsafe acts or conditions can cost a lot of money and result in a disabling injury, damage to equipment, and loss of property. Initial planning for the job is a must to avoid any unsafe acts related to working at heights.

Recommendations

For the benefit of employer's assets and welfare of the workers, the following proactive measures are recommended for safety protection to mitigate the identified hazards when working on elevated areas:

a. Provide the right equipment such as fall protection, temporary platforms, and man lifts suitable for the job, eliminating risks to worker safety, to keep them from taking chances or using shortcuts to perform assigned tasks.
b. Any elevated working area taller than 1.8 metres or six feet should be provided with a complete, properly working platform and guard rail system. A full body harness is required whenever a guard rail system cannot be installed.
c. Erect a scaffold with a complete working platform to any elevated working area above 1.6 metres.
d. Provide a man lift with a trained operator if the area is accessible and work can safely be performed on its operation.

Unsafe Acts of Workers Standing in Modified Man Basket

Heavy equipment is very useful in construction work. It helps to easily transport materials from one place to another and expedites work at the site which cannot be done manually. Using heavy equipment at site requires more attention, precautions, and competency of the operators. Most of the incidents related to heavy equipment are due to human behaviour leading to error. Employees tend to employ shortcuts because of a lack of resources or absence of the right equipment for the specific job when the timetable of the project has started to slip behind. Improper use of heavy equipment for other than its intended purposes can put workers in danger of serious injury or disability. Initial planning of any construction activities in the projects is essential and it should be considered prior to start-up. Failure to address these considerations at the early stage of the projects may force workers to cut corners, resulting in unsatisfactory work performance and unsafe acts and conditions. Every work discipline involved in the project should be part of the initial planning so they can provide the required equipment, manpower, and materials for the construction. Adequate equipment required for the project has to be identified prior to commencing any of the construction work.

Adequate equipment and resources shall be identified. Aside from equipment preparation, the company should also prepare certified operators with driving licenses to operate such equipment at the jobsite. The qualifications and driving licenses of the operators shall be specific to the heavy equipment to be operated. Additional heavy equipment operator's requirements such as training and certifications by a third party company authorized to issue a certification may also be required by the client. The qualifications and competency of the heavy equipment operators are the most important factors for safe operation of the equipment and meeting the construction schedule. It help to expedite to finish the assigned task and minimizing short cuts that leads to unsafe work practices as shown in the picture.

Recommendations

a. Conduct training for all employees involved in working at heights.
b. Site supervisor or foremen should conduct toolbox meeting before the work begins and address the proper working platform.
c. Ensure that every operator is competent to operate the equipment.
d. Verify if heavy equipment operator possesses a valid certification and driving license.
e. A written procedure of working at heights should be submitted to safety department for review.
f. Working at heights job safety analysis shall be reviewed and approved.

Unsafe Acts of Using a Stepladder

A stepladder is a type of ladder that is foldable and portable. It has two pairs of legs hinged on top with a spreader to lock the legs in place to form an inverted V. It can be made of aluminium, fibreglass, or wood, depending on the type of ladder required for a job to access an elevated working area. Because of its wide use, accidents involving a stepladder are common. Falls from ladders, as well as the ladder tipping over or collapsing, are the common incidents which may cause a serious injury and even death. Other hazards such as electrocution may happen if the incorrect type of stepladder is used near live electrical substations. The unsafe acts and conditions as shown in the picture indicated that incident is waiting to happen.

Dr. Pedro P. Marfa, PhD, MBA, BSCE, OSH Consultant

Recommendations

a. The user should never stand on the top of a stepladder to avoid tipping over the ladder or falling from it.
b. The employee should always check the condition of a stepladder prior to its use.
c. The employer shall provide appropriate ladders suitable to the working environment of the workers.
d. The employee shall only use the right-sized ladder at the elevated work. Using the right height of stepladder can prevent overreaching by the worker attempting to stand on the last rung of a stepladder, risking loss of balance and falls.
e. A buddy system is required, with a second working holding a stepladder on which a worker is standing, to prevent falls.
f. Barricade the area to prevent unnecessary personnel walking around the working area.

Unsafe Acts/Conditions in Excavation

Performing excavations is a must in any construction work. It is the first thing to do to build any structure in the building. Identifying and understanding the potential hazards in excavation is important during the development of a hazard identification plan (HIP) as prerequisite to fulfilling safety requirements. A comprehensive hazard identification plan has to be provided to construction personnel so they will be aware of potential hazards, prepare the heavy equipment and personal protective equipment, and ensure that an adequate number of trained workers are available to perform excavation activity.

Failure to comply with safety requirements in excavations may result in unsafe acts and conditions that can lead even to loss of life. A picture below shows serious hazards that require immediate action. The side wall of the excavation is ready to collapse, and it seems than none of the workers has realized that when that happens, all of them will be trapped and may even die. Occupational Safety and Health Administration (OSHA) standards for the construction industry 29 CFR 1926, Subpart P–Excavations outlines the requirements for a shoring protection that should be applied in the conditions shown in the picture. In addition to the unsafe acts and conditions observed in the picture, the worker inside the excavation has no access and no shoring protection, tailings remain near the edge of the excavation, and probably atmospheric testing at this depth of excavation was not enforced. These observations are common in construction sites where most of the people are not paying attention of their own protection at the beginning of work until incidents may have happened.

Once an incident has occurred, the contractor site management will provide immediate corrective actions. This passive action won't help protect the workers on site. Instead, a proactive approach to safety should always take place first in every aspect of the work. Another consideration to note is the competence of the workers. Every activity requires certain skills, and workers should undergo proper training to be familiar with the potential hazards associated with that activity and evaluate the risks involved. The management must be involved to ensure that workers are properly trained to perform their tasks in such a way as to mitigate unsafe acts and conditions at the construction site and prevent any incident in the confined work area. Adequate supervision must be available at the construction site and must continuously monitor the activities to improve the working behaviour of the workers.

Any construction activity in the confined space requires special procedures to address the roles and responsibilities of individuals involved with a confined entry activity, such as:

- Training for using atmospheric testing
- Supervisor responsibilities
- Standby man
- Entrants and rescue teams

Without proper training, the workers must not be allowed to enter into the confined work space.

Recommendations

    a. Assign a full-time employee to receive the work permit if the owner of the facility has the existing program of permit to work system (PTW).
    b. Provide proper shoring protection of the excavation, and provide an access ladder for their access/egress to climb out of the excavation.
    c. Conduct atmospheric testing every two hours.
    d. Assign a trained standby man in the confined space entry activity.
    e. Provide a log sheet to the standby man to record the entry and exit of the workers in the confined space.
    f. Provide proper training of the workers involved with the confined space entry activity and cement pouring activity.

Develop a hazard identification plan of any activities that involve equipment when working in the confined space entry. Guidelines on how to develop a hazard identification plan are outlined in chapter 8.

Dr. Pedro P. Marfa, PhD, MBA, BSCE, OSH Consultant

Unsafe Conditions of Power Hand Tools

Hand and power tools are the craftsman's best friend in the construction site. These tools are the bread and butter of a carpenter in day-to-day activities. It is essential that tools shall be regularly inspected prior to the start of a busy workday, and they must be cleaned after work upon returning them to the tool room. Misuse of these tools as well as wear and tear are the major causes of damage which could lead to user injury or even harm to co-workers at the construction site. Many carpenters at the construction site are not paying attention, even though they may have noticed defects in the hand tools as shown in the pictures. They keep using them until it results in injury or illnesses of the workers. These practices are a major concern of almost all construction companies.

The site supervisor and foremen have a major role in controlling workers' use of defective hand tools at the worksite. Identifying the unsafe conditions of hand tools will prevent potential personnel injury. A proactive approach of providing training on proper use and maintenance of hand tools supports improvement of employees' working behaviour. The training shall be done regularly as required by the construction company's occupational health and safety program.

To promote healthy working environments, any observed unsafe condition should be discussed by the site supervisor or foremen with all the employees during the weekly or daily safety meeting. Visual presentations using the defective hand tools will help workers understand the potential hazards and consequences. It also motivates them to inspect their hand tools before and after working hours.

To effectively utilize the hand tools at the construction sites, the following safety recommendations shall apply:

Recommendations

a. Appropriate personal protective equipment (PPE) shall be worn at all times when using hand tools.
b. Personnel shall not operate any tools unless they are appropriately trained in those tools' selection, use, inspection, and storage.
c. Grinding machines must be protected with guards. Those with defective or missing guards must be removed from the jobsite.
d. Power tools must be rated with 110V and protected with ground fault circuit interrupters (GFCI).
e. Extension cords must be double insulated with a three-prong plug.
f. Portable power tools such as grinding machines and drills shall be fitted with dead-man switches.

g. Tools constructed of good quality materials shall be used. Use of "home-made" tools is prohibited.
h. Tools shall be kept clean at all times.
i. Tools shall be inspected before and after use, as well as before storage.
j. Excessively worn, defective, or deformed tools shall not be used. If excessive wear, defect, or damage is observed, the tool shall be immediately tagged and withdrawn from use for repair or disposal.
k. Proper racks and boxes shall be provided and used for storage of tools.

Chapter 6

Hierarchy of Hazards Control

This chapter outlines the best safety practices and strategy for controlling potential hazards that will be encountered in the construction site. It helps construction contractor's management to identify the risk during the design stage and execution of the project. Each risk shall be assessed, evaluated, and prioritized to determine which control measures shall apply.

Hierarchy of hazard control is a system used in construction industry to minimize or eliminate exposure to hazards. It is a widely accepted system promoted by numerous safety organizations. This concept is taught to managers in construction industry, to be promoted as standard practice in the construction site.

The hazard controls in the hierarchy, in order of decreasing effectiveness, are the following:

- Elimination
- Substitution
- Engineered Controls
- Temporary Working Platforms
- Administration Control
- Personal Protective Equipment (PPE)

Elimination

The most effective hazard control at the construction site is to eliminate the hazard physically.

For example, if employees must work high above the ground, the hazard can be eliminated by moving the piece they are working on to ground level to eliminate the need to work at heights.

Substitution

The second most effective hazard control is substitution. It involves replacing something that produces a hazard (similar to elimination) with something that does not produce a hazard.

One example would be replacing lead-based paint with acrylic paint. To be an effective control, the new product must not produce other hazards. Because airborne dust can be hazardous, a product producing small-measure dust can be replaced by another that produces larger particles.

Engineered Controls

The third most effective means of controlling hazards is engineered controls. These do not eliminate hazards, but rather isolate people from hazards. Capital costs of engineered controls tend to be higher than less effective controls in the hierarchy; however, they may reduce future costs.

The engineering control method shall apply during the initial stage of the project through the design. It is essential that during the design phase, all the parameters of the project are included based on the project scope studies. It has to be cost effective and safe for the people, existing facilities, and the environment. A good risk assessment study of the project will help safety personnel develop a safety plan that meets these requirements.

Engineered controls apply during the engineering design stage of construction projects. Most common problems are encountered during implementation of projects where construction companies will tend to resort to shortcuts to perform certain activities at the jobsite. The probable justification for this is a lack of resources to complete the job, pressuring workers to function in a hazardous environment without proper protection. Unsafe acts and conditions are most likely to happen when companies are not proactive in the preparation required to complete the job. Examples of unsafe acts and conditions that could occur at the construction site.

Many best safety practices at construction sites apply engineered controls to provide worker protection. How do they work, and when will they apply? You can find answers in the following pages in the given cases.

Temporary Working Platforms

Health and safety protection of the workers are the primary concern of construction companies. Without them nothing will happen on the construction site. In other words, the workers are the most precious assets in the company, and there is no substitute if one of them meets with an accident or death. Protection of these workers is imperative and requires attention from the company to protect them from falls in elevated working areas. Any situation such as like this should be analysed by identifying the potential hazards associated with working at heights. The hierarchy of controls may the best tools to apply whenever the construction activities involve height work.

How effective are engineered controls in providing a safe working environment at the worksite? Let's consider an example to understand how effective engineering control methods can be at the worksite.

Example: Four masonry workers were tasked to lay concrete masonry blocks to a twenty-four metres long by ten metres high vertical wall of a material warehouse building. The site supervisor instructed the masonry workers to complete the block work within one week, since the target completion of the project had been delayed. The masonry workers were given no choice and feared to ask the site supervisor about the materials to be used for a temporary working platform to be employed when they reached the height where they would be unable to reach the next level of blocks. On their own initiative, when they reached a height of five feet, the workers began to stack the masonry blocks on the ground and lay wooden planks on them. Standing on this platform enabled them to left the reinforcing rebar and insert it through the masonry blocks cells for the next level. When they had reached the height of seven feet, the client's safety representative was conducting a site safety inspection and noticed the unsafe conditions of the temporary platform. He temporarily halted work, reported his observation to the site supervisor, and instructed him to take immediate corrective action.

In this situation, which among the hierarchy of controls is appropriate to correct the conditions? I submit that engineered controls are the appropriate action to correct the site conditions. A proper scaffold, able to carry the weight of the workers should be erected alongside the wall as a temporary working platform to provide them with a safe working area for laying the concrete masonry blocks. A guard rail system should be provided around the edge of the temporary working platform with proper access and egress of masonry workers. Providing a proper working platform in the elevated area is considered as an engineered control method.

Example: In an oil and gas producing facility, a pipe-fitter and welder were tasked to weld a three-metre length of twelve inch diameter pipe spool, fitted with a ninety degree elbow, at both ends of an existing hydrocarbon pipeline on the pipe rack support. The area was open and accessible for the crane to assist a lift and hold the pipes while the pipe-fitter and welder completed the welds. Since the pipe spool had to be looped alongside the pipe rack, about 2.5 meters away from the pipe support, the pipe-fitters needed a scaffolding platform to enable them to perform the operation. For them to complete the task, a temporary working platform needed to be erected to cover the loops where the pipe spool has to be joined with the existing hydrocarbon pipeline.

Of course the construction company should provide the temporary platform for the pipe-fitter to carry out the welding work. However, it takes a lot of time to erect the temporary working platform, and it may be costly for them due to a volume of scaffolding to be used of a platform.

Erecting a temporary working platform all over the area to cover the pipe loop is an acceptable measure for completing the process of connecting the pipe spool with elbows at both ends by assisting the crane. However, if the fabrication of the pipe spool assembly will be done on the ground and raised by the crane as soon as it is completed, it would be safer and easier to assemble on both sides of the joints rather than lift them in pieces and join them to the existing line.

Both ways are acceptable with the same results in terms of performing the job to connect the spool, but for the safety and access of the workers, the best approach is to assemble it on the ground, minimizing the potential exposure of the pipe-fitter working at height to hazardous conditions. This is an example of engineered controls.

Administrative Controls

Administrative controls are changes to the way people work. Examples of administrative controls include procedure changes, employee training, and rotation of work shifts; starting work earlier in the morning during summer; and installation of signs and warning labels (such as those in the Workplace Hazardous Materials Information System or WHMIS). Administrative controls do not remove hazards, but limit or prevent workers' exposure to the hazards, such as completing road construction at night when fewer people are driving.

Personal Protective Equipment (PPE)

Personal protective equipment (PPE) includes gloves, respirators, hard hats, safety shoes, safety glasses, high visibility clothing, and safety footwear. PPE is the least effective means of controlling hazards because of the high potential for damage, which can render PPE ineffective. Additionally, some PPE, such as respirators, increase the physiological effort needed to complete a task and therefore may require medical examinations to ensure workers can use the PPE without risking their health.

Personal protective equipment is the last line of defence to protect the workers from harm due to construction hazards. It is the most economical safety protection measure to provide the workers for fulfilling their daily tasks at the construction site.

I want to share with you the tiny personal protective equipment that you can use in a noisy environment: earplugs. Most workers overlook the benefits of using such personal protective equipment as tiny earplugs, unfashionable glasses, heavy headgear, and bulky safety shoes. Comments from workers at a construction site give common excuses for their not wearing personal protective equipment at work: they feel uneasy when wearing safety shoes because they are heavy, dizzy when wearing safety glasses, uncomfortable when wearing a safety helmet instead of a baseball cap; and so forth. Some of them, when instructed to wear PPE, respond to the safety officer that they have been working for many years on the same job and routine, and nothing has happened to them. Most of them assume that what has not happened yesterday will not happen in the future.

Some workers at the construction site say they are wearing the personal protective equipment only because it is company policy, and they don't want to be reprimanded if a company safety officer catches them not wearing the required safety protection. Other workers say they are not used to wearing such types of eye protection since they wore fashion glasses in previous projects. Dealing with this personal behaviour at the construction site makes a safety officer's job difficult and affects the company's reputation. The safety officers sometimes are frustrated to find ways to enforce and educate workers to wear the required safety gear to protect them from untoward incidents in the course of doing their jobs to meet the target schedule. In other instances, workers challenge the company safety officer by saying, "I have been here in construction work for many years and never been involved with an incident; look, I'm still healthy and fine." This ridiculous response from workers poses a big challenge to safety practitioners in improving workers' safety awareness, their working attitude, and their behaviour on the job.

As the author of this book, I want safety personnel to be ready for the challenges ahead and to equip themselves with knowledge that will influence them to work safely. The examples I supply are real situations that have happened in the construction field. Understanding the behaviour of workers at the jobsite will help you succeed in improving their level of understanding and awareness of safety when performing their jobs at the construction site.

As a company's safety officer assigned to the project to implement the occupational safety and health program in the construction site, your first step should be to conduct an assessment the level of safety awareness of all the employees working at site. Based on the outcome of the assessment, you will be able to address the weakness of the workers and improve their safety awareness for their own protection. The working behaviour of each worker should be monitored, and any observed unsafe behaviour must be corrected.

Although the company safety officer will closely monitor employees' compliance with the company policy of wearing personal protective equipment while working at the construction site, many workers may still refuse to do so. To boost the morale of all employees and motivate them to follow the company policy, the workers should understand that the employer or company is doing its best to provide a safe working environment and investing a huge amount to purchase personal protective equipment for their protection.

Workers must also realize that the company priority is to protect employees from dynamic hazards associated with construction work and that the company is committed to provide them with resources for their protection. Likewise, the workers should realize that they are the beneficiaries of wearing the personal protective equipment and following the company's policy. They should understand the situation, bearing in mind that any incident involving a minor or major injury affects not just the workers but also their family and loved ones. It may not at once change their working behaviour in the field, but in the long run they will understand the value of the personal protective equipment for day-to-day activity at the workplace. The learning curve on this may notice at the middle of peak period of construction.

Continuing education through safety training and meetings for the workers is the most effective way to support them by widening their understanding of safety. Employees must be motivated to follow safety procedures on their own without being told by anyone to follow the company policy.

The basic personal protective equipment that workers normally wear every day at the construction site are the following:

- Eye and face protection shall meet with the American National Standards Institute (ANSI, Z87.1-1968) as per 29 CFR Part 1926, Occupational Safety and Health Administration (OSHA) standards for the construction industry.
- Head protection/safety helmets against impact and penetration of falling objects shall meet with ANSI (Z89.1-1969).
- Head protection/safety helmets against a high-voltage electrical shock shall meet with ANSI (Z89.2-1971).
- Safety shoes/foot protection shall meet with the standard requirements ANSI (41.1-1967).

Specialized personal protective equipment for specific jobs such as respiratory protection, fall protection, electrical gloves, etc., shall also meet ANSI standards.

Chapter 7

How to Conduct Safety Meetings Effectively

This chapter discusses best practices when conducting safety meetings with workers at the construction site. It will enable site supervisor and safety officers to help workers understand the topics and associated potential hazards to be discussed in the meeting.

Chapter Objectives

After reading this chapter, you should know the following techniques for conducting effective safety meetings:

- Describe the potential hazards of selected topics to be presented in the safety meeting.
- Identify the potential hazards discussed in this chapter and explain the corrective actions.
- Identify the corrective action for each hazard presented in this chapter.
- Illustrate the best safety practices by showing visual aids to the workers and explaining the benefits of working safely.
- Enumerate the corrective action of each hazard presented in the meeting.

Recommended best safety practices of conducting safety meetings with workers should follow three basics steps:

- Step 1: Explain the potential hazards.
- Step 2: Identify corrective actions and controls.
- Step 3: Provide visual aids and demonstration.

Safety Topic 1: Eye Protection

Step 1: Explain the Potential Hazards

Many construction workers perform their daily jobs at the construction site without wearing eye protection. Just think of the eye hazards in our workplace:

- Flying objects, dust and foreign bodies
- Welding arcs
- Arc flash
- Sparks and slag from welding and brazing works
- Abrasive from grit blasting
- Chemical splash

Failure to protect our eyes could leave us with limited sight or none at all.

Step 2: Identify Corrective Actions and Controls

We only have one pair of eyes and we must protect it at all times. Wearing the right protection can prevent most eye injuries. Follow these basic tips:

- Don't wear contact lenses on site. Dust and other particles can get under the lens.
- Keep your safety glasses on when you wear other protection such as a welding helmet or face shield. Why? Because when you lift up a visor or shield, you may still be exposed to flying objects, dust, or other hazards at the construction site.

- Match your eye protection to the hazard. Goggles that protect you from dust may not protect you from splash or radiation.
- Make sure your eyewear fits properly.
- Clean your lenses with water or a lens cleaning solution to float the dirt away instead of scratching it into the lenses.
- Get your eyes checked regularly to make sure that no problem has developed.

Step 3: Provide Visual Aids and Demonstration

Take a look at eye protection used by your co-workers, and point out any cracks or damage to the workers.

Review the company policy on providing and replacing eye protection.

Review any special requirements for welding masks, sandblasting hoods, face shields, etc. This will help workers understand the benefits of using eye protection.

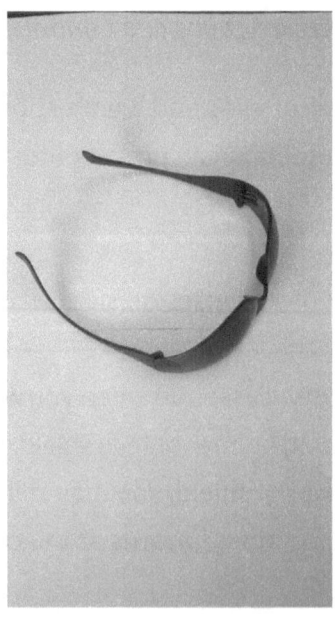

Safety Topic 2: Hearing Protection

Step 1: Explain the Potential Hazards

Many workers are overexposed to noise. In time, overexposure can damage your hearing. Hearing loss prevents you from hearing other hazards on the jobsite. It also causes problem in your personal life.

- It interferes with how you hear normal speech.
- It prevents you from socializing.
- It prevents you from listening a good music.
- It may cause high blood pressure if you become upset because you cannot hear what your co-workers are saying.
- It can become permanent.

Step 2: Identify Corrective Actions and Controls

Hearing loss is preventable by using ear protection when working in a noisy area or using equipment which produces a loud noise and vibration.

Noise is any unwanted sound. There are two types: continuous noise (e.g., turbines) and impulse noise (e.g., jackhammers).

Noise is measured in decibels (dB). For example, a quick cut saw produces 115 decibels, a jackhammer 110 decibels, and a drill machine 100 decibels.

Think about this: when the noise level is eighty decibels and it goes up to eighty-three, the noise is twice as loud. In the same way, the noise level drops three decibels when you double your distance away from this activity. Without hearing protection, your safe noise level for an eight hour day with no other noise exposure is eighty-five decibels. This is the loudness of a room full of people. When the noise cannot be reduced or controlled and exceeds ninety decibels, you need to wear hearing protection.

Step 3: Provide Visual Aids and Demonstration

Identify tasks on site that require hearing protection.

Review the company policy and procedures regarding hearing protection, and show them the two types of hearing protection:

- Earplugs
- Earmuffs

If possible show them how to use and insert the earplugs.

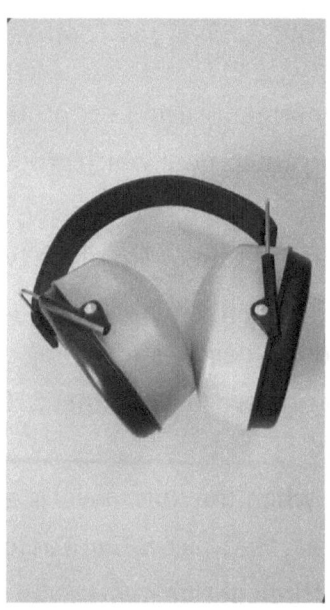

Safety Topic 3: Hand Protection

Step 1: Explain the Potential Hazards

The best tools we have are our hands, thus we need to protect them in our day to day activity on the job. Manual work expose our hands to many different hazards, from cuts to chemical, from pinching to crushing, and from blisters to burns if we are not paying attention to protect from any of these hazards.

Step 2: Identify Corrective Actions and Controls

Leather gloves provide good protection against sharp objects, splinters, and heat. Cotton or other fabrics don't stand up well to these conditions. You should wear them only for light duty jobs. Wearing anti-vibration gloves when using power tools and equipment can help prevent hand-arm vibration syndrome.

Hand-arm vibration syndrome causes the following changes in fingers and hands:

- Circulation problems such as whitening or discoloration, especially after exposure to cold.
- Sensory problems such as numbness and tingling.
- Musculoskeletal problems such as difficulty with fine motor movements like picking up small objects.

Our hands also need protection against chemicals. Check the label to see whether a product must be handled with gloves and what types of gloves are required. If that information is not on the label, check the Material Safety Data Sheet (MSDS).

Using the right gloves for the job is important. For example, rubber gloves are no good with solvents and degreasers. The gloves will dissolve on contact.

Step 3: Provide Visual Aids and Demonstration

Talk about the specific chemicals used on your jobsite and the type of gloves recommended for each activity.

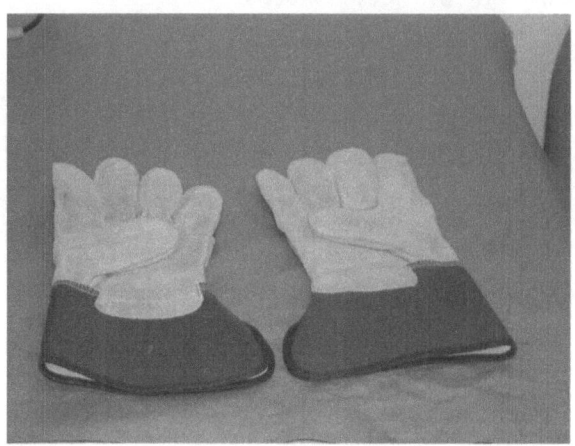

Safety Topic 4: Fall Protection

Step 1: Explain the Potential Hazards

Falls are the number one cause of accidental deaths in the construction industry. And you don't have to fall far to be killed.

Step 2: Identify Corrective Actions and Controls

On many construction sites, guard rails are the most common and convenient means of fall protection. Where guard rails cannot be installed or are impractical, the two types of fall protection are travel restraint and fall arrest systems. Both involve a full body harness.

A *travel restraint system* keeps you from getting too close to unprotected edges, such as roof eaves and roof deck. The lifeline and lanyard are adjusted to let you travel only so far. When you get to the open edge of a floor or roof, the system holds you back and prevents you from falling. A full body harness should be used with a travel restraint system. You can attach the harness directly to a rope grab on the lifeline or by lanyard. The lifeline must be securely anchored.

If no other fall protection is in place, you must use a *fall arrest system* if you are in danger of falling

- into operating machinery,
- into water or another liquid, or
- into a hazardous substance or object.

A fall arrest system consists of a full body harness, a lanyard, and a shock absorber lanyard. You can connect the lanyard directly to adequate support or to a rope grab mounted on an adequately anchored lifeline. A full body harness must also be worn and tied off when you are

- on incomplete scaffold platform with our guard rail system at 1.8 meters tall or
- getting on, working from, or getting off a suspended platform or suspended scaffold.

The lifeline must be adequately or properly anchored to a structure that can withstand at least five thousand pounds.

Step 3: Provide Visual Aids and Demonstration

Show how to put on, adjust, and wear a full body harness.

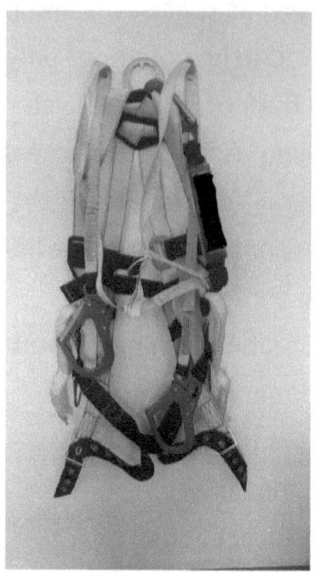

Safety Topic 5: Stepladders

Step 1: Explain the Potential Hazards

The stepladder is one of the most familiar items on a jobsite. Still, workers get hurt using them. Falls are the biggest risk. Even though workers are not very high off the ground, some have died from failing a short distance and landing the wrong way. Even sprains or strains could mean pain and days off work.

Step 2: Identify Corrective Actions and Controls

Use a stepladder to demonstrate the following points in your talk. Here's how to use a stepladder the right way:

- Check the ladder for defects or damage, at the start of your shift, after it has been used elsewhere by other workers, and after it has been left in one place for a long period of time.
- Keep the area at the base of the ladder clear from any obstructions.
- Make sure the spreader arms lock securely in the open position.
- Stand no higher than the second step or rung from the top.
- Never use the stepladder as a walkway.
- When standing on the ladder, avoid leaning forward, backward, or to either side.
- Always open the ladder fully before using it. Don't use an unopened stepladder as a straight ladder or as an extension ladder.
- Never stand on the top step, the top, or the pail shelf of a stepladder.
- When climbing up or down a stepladder, always face the ladder, and maintain three-point contact. This means that two hands and one foot or two feet and one hand must be on the ladder at all times.

Step 3: Provide Visual Aids and Demonstration

Inspect stepladders in use on site, and determine whether other equipment would provide safer, more efficient access.

Dr. Pedro P. Marfa, PhD, MBA, BSCE, OSH Consultant

Safety Topic 6: Extension Ladders

Step 1: Explain the Potential Hazards

Extension ladders can be dangerous tools if improperly used. Workers have been killed and injured from falls and power line contact.

Step 2: Identify Corrective Actions and Controls

- Choose the right ladder for the job. It must be long enough to
    a. Be set up safely at a seventy-five degree angle and
    b. Extend at least ninety centimetres (three feet) beyond the top landing.

- A two section extension ladder should be no longer than fifteen metres (fifty feet); a three section ladder no longer than twenty metres (sixty-six feet).
- Check the ladder for damage or defects before you set it up, after it has been used elsewhere by other workers, and after it has been left somewhere for a long period of time.
- Set the ladder on a firm, level base. If the base is made of soft, loose, or wet material, clear it away, or stand the ladder on a mud sill.
- Never erect extension ladders on boxes, carts, tables, or other unstable objects. Never stand them up against flexible or movable surfaces.
- Set the ladder up at a safe angle one foot out for every four feet up depending on the length.
- When the ladder is fully extended, sections must overlap at least ninety centimetres (three feet).
- Tie off or otherwise secure the top and bottom of the ladder. Keep areas at the top and bottom clear of debris, scrap materials, and other obstructions.
- When climbing up and down, always face the ladder and maintain three-point contact.
- Don't carry tools, equipment, or materials in your hands while climbing. Use a hoist line or gin wheel for lifting and lowering materials.
- Be very careful when erecting extension ladders near live overhead power lines. Never use a metal or metal-reinforced ladder near electrical wires or equipment.
- When you must work from a ladder more than three meters in height, wear a safety harness, and tie off to a well anchored lifeline or other support to the ladder.

Step 3: Provide Visual Aids and Demonstration

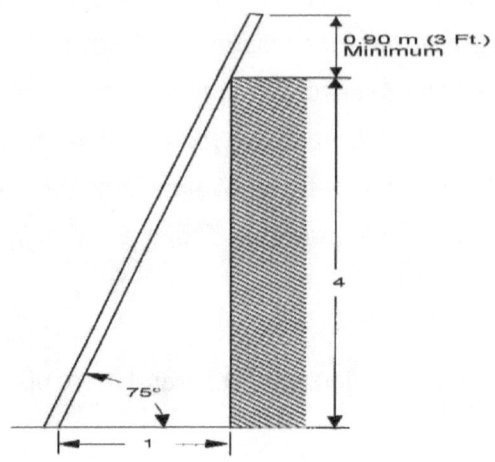

Safety Topic 7: Electrical Safety

Step 1: Explain the Potential Hazards

Using electricity on site can be hazardous in three areas especially:

   a. Tools
   b. Cords
   c. Panels/generators

Step 2: Identify Corrective Actions and Controls

Consider all electrical wires and equipment energized until they are tested and proven otherwise. No worker shall connect, maintain, or modify electrical equipment or installations unless

a. the workers are trained and certified as electricians or
b. the workers are otherwise permitted to connect and maintain or modify electrical equipment or installations under direct supervision by a qualified and competent person.

A worker who does not meet the requirements of (a) or (b) may only insert or remove an electrical attachment plug of electrical equipment to or from power receptacles.

a. Tools

- Use only electrical tools that have Underwriters Laboratories (UL) or equivalent certification.
- Make sure the casings of double insulated tools are not cracked or broken.
- Any shock or defect, no matter how small, means that the tool or equipment needs to be checked and repaired.
- Take defective tools out of service.
- Always use a ground fault circuit interrupter (GFCI) with portable electrical tools operated outdoors or in damp or wet locations. GFCIs detect current leaking to ground from a tool or cord and shut off power before damage or injury can occur.
- Before drilling, nailing, cutting, or sawing into walls, ceilings, and floors, check electrical wires or equipment conditions.

b. Cords

- Make sure that tool cords, extension cords, and plugs are in good condition.
- Use only three-pronged extension cords.

- Make sure that extension cords are the right gauge for the job to prevent overheating, voltage drops, and tool burnout. A twelve-gauge extension cord is ideal.
- Do not use cords that are defective or have been improperly repaired.
- Protect cords from traffic to prevent them from being damaged.
- When outdoors or in wet locations, plug into a GFCI-protected receptacle, or use a portable in-line GFCI.

c. Panels/Generators

- Temporary panel boards must be securely mounted in a lockable enclosure protected from weather and water. The panel boards must be accessible to workers and kept clear of obstructions.
- Receptacles must be GFCI protected.
- Use a portable generator with built-in GFCI receptacles, or use a portable in-line GFCI at the generator receptacle.

Step 3: Provide Visual Aids and Demonstration

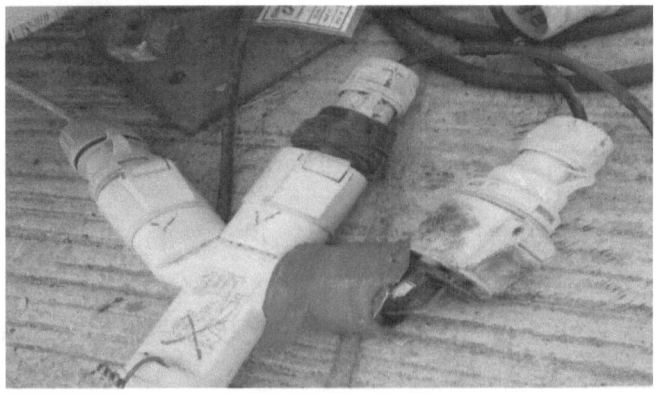

Safety Topic 8: Trenching and Excavation Inspection

Step 1: Explain the Potential Hazards

Without regular and frequent inspection, you have no assurance that sloping, shoring, or trench boxes are effective in protecting workers from trench collapse.

Step 2: Identify Corrective Actions and Controls

Sloping, shoring, and trench boxes must be inspected regularly. Inspection is everyone's responsibility.

With hydraulic shoring, look for:

a. Leaks in hoses and cylinders
b. Bent bases
c. Broken or cracked nipples
d. Cracked, split, or broken sheathing

Report any of these conditions to your supervisor.

Check timber shoring before it's installed. Discard any damaged or defective lumber. With timber shoring, check for:

a. Cracked or bowed sheathing
b. Loose or missing cleats
c. Split or bowed lumber
d. Damaged bracings and off level

If the wooden lumber shows signs of crushing, this indicates structural problems and the need of more struts or supports. Make sure that shoring members are the size required by regulations for the depth of your trench and type of soil.

Always check areas near shoring where water may have seeped in. The combination of water and granular soil can lead to washout. This undermines the trench wall and has killed and injured workers many times in the past.

In trench boxes, look for:

a. Damage and other defects
b. Deformed plates
c. Cracks in weld
d. Bent or distorted welds in sleeves and struts
e. Missing struts
f. Bent struts
g. Holes, bends, or other damage to plates

During use, check the trench box regularly and often to make sure that it is not shifting or settling more on one side than the other. This can indicate movement of soil or water underneath. If the box is shifting or settling, get out, and tell your supervisor about it.

Ground trenches should be inspected for tension cracks. These may develop parallel to the trench at a distance of about one half to three quarters of the trench depth. If you find cracks in the ground, alert the workers, and double-check the shoring or trench box. It's dangerous to overlook damage or defects in protective systems. Even though the job is short term or almost finished, trenches can still cave in.

Step 3: Provide Visual Aid and Demonstration

Inspect sloping, shoring, and trench boxes on site, and check ground conditions nearby.

Safety Topic 9: Confined Spaces – Dangerous Atmosphere

Step 1: Explain the Potential Hazards

Dangerous atmosphere has killed those working in confined spaces as well as those attempting to rescue them. You should know the hazards.

Step 2: Identify Corrective Actions and Controls

Dangerous types of atmosphere are the following:

- Flammable and explosive
- Toxic
- Oxygen deficient
- Oxygen enriched

a. *Flammable and Explosive*

- Natural gas from leaking gas lines or natural sources
- Methane from decaying sewage
- Propane gas from leaking tanks and spills
- Vapours from solvents used for painting, cleaning, and refinishing

b. *Toxic*

- Vapour from solvents
- Hydrogen sulphide from decaying sewage or raw petroleum
- Carbon monoxide from engine exhaust

c. *Oxygen-deficient* atmosphere contains less than 19.5 per cent oxygen. Breathing oxygen-deficient air can make you lose judgement, coordination, and consciousness. In a confined space, oxygen can be displaced by other gases or used up by rusting metal, combustion, or bacteria digesting sewage.

d. *Oxygen enriched* atmosphere contains more than twenty-three per cent oxygen. This is rare and usually caused by leaking oxygen hoses or cylinders.

Check for atmospheric hazards before entering any confined space. Use properly calibrated gas detection equipment. Many dangerous atmospheres cannot be detected by smell or taste. Make sure the equipment is able to detect what you suspect. Some detectors have sensors that check for oxygen content, explosive gases or vapour, and a range of toxic gases. Some have only one or two sensors and may not detect certain types of hazards. You may need more than one detector; few detectors can test for everything.

Check all levels of the space. Some contaminants are lighter than air and accumulate near the top of the space. Others are heavier than air and settle at the bottom of the space. If you leave the space for a break or for lunch, test before you go back in. Dangerous atmospheres can develop without warning. If tests indicate a dangerous atmosphere, you must not enter the space until it is thoroughly ventilated and subsequent tests indicate the air is safe for workers.

Ventilation and testing must continue as long as you are in the space. If the space can't be adequately ventilated, you can only enter if:

- You are equipped with suitable respiratory protection and a full body harness, attached to a rope anchored outside the space and held by a standby man with an alarm;
- You have a means of communication with the standby man outside the confined space; and
- A person trained and equipped in artificial respiration and emergency rescue is available outside the confined space.

Never attempt to rescue a worker overcome in confined space unless you are trained and equipped for it. Many workers trying to save their buddies have become victims themselves. Get emergency help.

Step 3: Provide Visual Aids and Demonstration

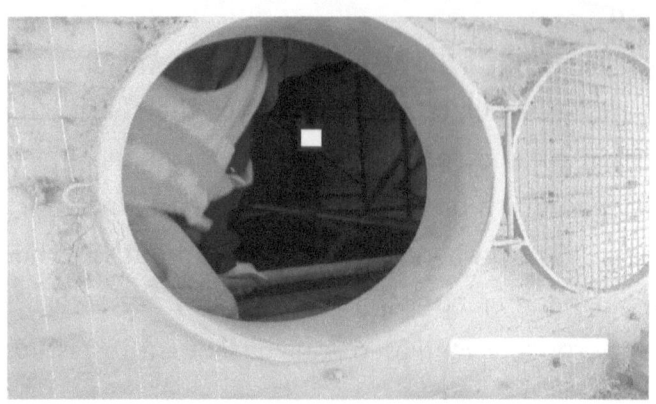

Safety Topic 10: Housekeeping

Step 1: Explain the Potential Hazards

Every year, poor housekeeping and storage account for a significant percentage of accidents and injuries in construction. We all know how fast garbage accumulates on site: scrap lumber, broken bricks, pieces of drywall, strap bands, pipe spools, surplus construction materials, etc. Construction rubbish is often irregular in shape, hard to handle, and full of sharp objects. One of the biggest problems is packaging. Too often it gets removed from material and left wherever if falls. This creates tripping and slipping hazards. It also makes other hazards hard to see. Even worse, it invites more mess. When a site isn't cleaned up, no one cares about leaving garbage where it drops. When that happens, you can't see faulty wiring, protruding nails, damaged flooring, and gaps in scaffold planking.

How can you concentrate on your work when you're worried about slipping, falling, or tripping over the garbage and debris? Production and installation time go up, while quality tails off. A messy environment also makes it difficult to use material handling equipment. As a result, more material gets handled manually, which increases the risk of injury and damage.

Step 2: Identify Corrective Actions and Controls

Housekeeping means cleaning up scrap and debris, putting it in containers, and making sure the containers are emptied regularly. It also means proper storage of materials and equipment.

Effective housekeeping and storage prevent accidents and injuries.

- Clean up as work progresses.
- Keep equipment and the areas around equipment free of scraps and debris.
- Keep stairways, ramps, and other travel areas clear.
- Secure loose or light material stored on roofs and open floors to keep it from blowing around in the wind.
- Never let material fall from any level of the project. Use an enclosed chute, or lower the material in containers.
- Keep material at least 1.8 meters or six feet away from floor and roof openings, floor and roof edges, and excavations and trenches.
- Store materials so that they won't roll or slide in the direction of an opening. Use blocking if necessary.
- Before handling used lumber, remove or bend over any protruding nails, and chip away hardened concrete.
- Remove flammable rubbish and debris immediately from the vicinity of welding, flame cutting, propane heating, and other ignition sources.

Dr. Pedro P. Marfa, PhD, MBA, BSCE, OSH Consultant

Step 3: Provide Visual Aids and Demonstration

Review housekeeping problems unique to your workers. Discuss housekeeping problems found in other areas on-site for everyone's awareness.

Chapter 8

Developing a Hazard Identification Plan (HIP)

The purpose of this chapter is to familiarize construction company project engineers, site supervisors, and safety staff with the process of developing a hazard identification plan (HIP) for the project prior to beginning any construction activity.

Chapter Objectives

After reading this chapter, you should be able to do the following:

- Describe the potential hazards identified during the site walk-through.
- Develop the hazard identification plan for each activity, and define the corrective action(s) for implementation.
- Identify the potential hazards associated with the scope of work.
- Recognize the hazards classification on each sample provided in this chapter.
- Identify the recommended corrective actions or safety measures for each identified hazard at the construction site.
- Recognize the steps of developing a hazard identification plan as provided by a sample in this chapter.

Developing a hazard identification plan is a process for identifying possible situations where people may be exposed to injury, illness, or disease that may result and describing how work is to organized and managed in view of those risks. It is essential to the company to develop a hazard identification plan prior to starting any of the construction activity. Each employee should understand the process of identifying the potential hazards in the project, based on the scope of work. This will help them eliminate or prevent exposure to injury and ensure protection of the company assets in the existing facility.

Construction company site management should be familiar with the process of developing a hazard identification plan and support the project engineer in developing the plan, with the assistance of the safety team, to provide recommendations in every potential hazard that may be encountered during the construction. Every discipline involved with the project should participate in this process since they are the experts in their field and are able to identify potential hazards based on the scope of work of the project. Some of the project engineers and company safety officers may not be familiar with the process, but the client's safety representative can help guide them in developing the plan.

At the bidding stage of the project, the construction company will develop a hazard identification plan that is based on the project scope provided by the client. This plan is to be submitted as part of the company occupational health and safety program for reference during the project bidding stage. As soon as the construction company is awarded the project, the client representative will advise the company that the specific safety program and a detailed hazard identification plan shall be submitted to the client's representative at most fifteen working days from the date of signing the contact, before issuance of the notice to proceed to the contractor as described in the contract agreement.

A team composed of the project engineer, facility owner, company safety officer, and client safety representative will conduct a site walk-through to identify the potential hazards of the project. All the information gathered during the walk-through shall be consolidated. Development of hazard identification plans may vary in different countries, but the essence of the hazard identification plan is similar. A structured hazard identification plan will help identify the corrective actions for each construction activity to prevent incidents and worker injury.

Since the prime objective of the company – to protect the workers from the potential hazard in the construction site – is the same also as the client perspective, safety is stressed during the job explanation meeting as a reference of the project owner. Construction companies normally will develop a hazard identification plan based on the project scope of work provided by the client. Below are the basic steps of developing the hazard identification plan based on the project scope of work.

How to Develop a Hazard Identification Plan (HIP)

The following basic steps in developing a hazard identification plan will help project engineers, site management personnel, and safety personnel. All you have to do is to write down the activities involved with the project and consolidate them in this order:

a. Activity involved,
b. Potential hazards on each activity, and
c. Safety measures or corrective actions. (Note: Giving safety measures on each potential hazard should not only focus on the hazard but consider the internal safety regulations and work procedures in the facility. It should be addressed in the HIP clearly so that all the workers at the jobsite will understand.)

Let's go over the examples of construction activities below and identify the hazards as an exercise to develop a structured hazard identification plan for your project.

Trenching and Excavation

List the potential hazards and conditions that could possibly trigger an incident in a trench and excavation area where injury of the workers may occur. Each hazard should be provided with corrective actions or recommendations for mitigating the hazard.

Potential Hazards and Conditions

  a. Presence of existing underground utilities such as power cables, communication cables, water lines, and sewer lines
  b. Vehicle traffic near excavation
  c. Unstable soil conditions
  d. Falls and trips into the open trench or excavation
  e. Improper placement of access and egress from excavation
  f. Placement of tailings at the edge of the trench and excavation
  g. Heavy equipment operation near the edge of excavation
  h. Defective hand tools

Corrective Safety Measures

  a. Presence of existing underground utilities

   - Obtain written permission from the project owner prior to start of any activity.
   - Locate power cables exactly by using a probe cable detector.

- Provide a protection of all underground utilities as soon as they are exposed.

b. Vehicle traffic near excavation

- Provide flagman with a coloured vest to direct the traffic.
- Install physical barricades around the trench or excavation.
- Provide flashing lights during work and at night to warn vehicles or pedestrians.

c. Unstable soil conditions

- Provide shoring protection at the edge of excavation or trench to prevent cave-in.
- Side slope the edge of the excavation to prevent erosion.

d. Falls and trips into the open trench or excavation

- Organize all the construction materials delivered at site.
- Perform housekeeping before the end of shift.
- Install warning signs and barricades around the excavation.
- Do not leave the trench or excavation open for a long period of time.

e. Improper placement of access and egress

- Secure a straight ladder in the trench to prevent falls.
- Extend the ladder to at least three rungs or 0.90 metre from the landing or at the edge of the excavation.
- Maintain a proper angle of a straight ladder provided as access of workers from the excavation to at least 1:4

ratio of distance of base to unit height or seventy-five-degree angle.

f. Placement of tailings at the edge of excavation

- Maintain clearance from the edge of excavation to any tailings of at least two feet.
- Collect the tailing materials away from the edge of excavation.

g. Heavy equipment operation

- Maintain a safe distance away from the edge of operation of at least 1.5 metres.
- Heavy equipment operator must possess a government-issued driving license.
- Provide a flagman to direct heavy equipment operation.
- Install backup alarm and beacon light on all heavy equipment working in the construction site. Backup alarm will help alert the workers around the area.

h. Defective hand tools

- Conduct regular inspection of all the hand tools to be used for excavation. Any defective tools must be removed immediately from the construction site.
- Clean the hand tools at the end of every workday, before they return to the tool room.
- Conduct training in proper use of hand tools.

Rebar Works

Rebar works are the most common activity in project construction in preparation for pouring concrete in flooring slabs, superstructure walls, columns, and foundations. Installation of rebar matting on the floor slab, pre-assembly prior to placing it onto the foundation, and casting in place to the foundations are normally done manually at site. The following considerations should be addressed in the hazard identification plan to prevent worker injury during the installation.

Potential Hazards and Conditions

   a. Trips and falls between spaces of rebar
   b. Protruding rebar
   c. Ergonomics

Corrective Safety Measures

   a. Trips and falls between spaces of rebar

   - Provide a piece of wood over the rebar for workers to step on during rebar assembly.
   - Use full body harness when tying off the rebar on an elevated area.
   - Conduct safety toolbox meeting prior to start of workdays.

   b. Protruding rebar

   - Provide caps for all protruding rebar to avoid impalement.
   - Install a physical barricade around the rebar assembly area.

c. Ergonomics

- Conduct safety toolbox meeting and discuss proper lifting procedures.

Painting in Confined Space

There are some jobs involved in the project that need to be done in confined space entry areas such as inside the vessel, sewage manhole, tanks, and excavation deeper than four feet. Depending on the nature of the job to be performed, the potential hazards must be identified and provide a corrective actions prior to begin any of the activity. The facility owner may require the contractor to provide a specific hazard identification plan before they will issue a permit to work inside the confined space.

Potential Hazards and Conditions

The potential hazards that may be encountered include but are not limited to the following;

a. Atmospheric hazards
b. Struck by, caught in between, and pinch points
c. Solvents

Corrective Safety Measures

a. Atmospheric hazards

- Obtain permission from the facility owner prior to starting any activity in the confined space.

- Conduct atmospheric test before sending workers into the confined space.
- Assign a full-time standby man while workers are in the confined space.
- Conduct confined space training of all workers involved in confined space entry works.
- Provide log sheet to register by name the entry and exit of all workers in the confined space.
- Provide air blowers and/or air extraction while workers are in the confined space.
- Train workers for a rescue team.
- Provide a standby self-contained breathing apparatus (SCBA) near the entrance to the confined space.
- Provide full body harness for each worker in the confined space and attach the lanyard to a lifeline.
- Provide an alarm to notify the workers in the confined space in case of emergency.
- The stand-by man should not leave the area until arrival on-site of a person to relieve him.
- Continuously monitor the oxygen level at least every two hours.

b. Struck by, caught in between, and pinch points

- Remove excess materials at the working area, and perform housekeeping on a daily basis.
- Provide protection for welding cables and extension cords passing through an open manhole cover, doors, and other objects that may cause damage to the cables.

c. Solvents

- Provide personal protective equipment such as goggles, safety glasses, gloves, and respirators.
- Respirator cartridges must be replaced regularly, and make sure that the cartridge addresses the chemical(s) coming from the paint(s) in use.
- Provide Material Safety Data Sheets (MSDS) to the site supervisor and painter(s) as their reference in case of emergency.

Welding Work

Potential Hazards and Conditions

The most common conditions that will trigger hazardous situations during welding works and must be addressed in the HIP are the following:

a. Heat exhaustion or heatstroke due to excessive heat during welding
b. Smoke produced from melting metals and welding rods
c. Sparks and noise produced from grinding machines
d. Slips, trips, and falls
e. Poor housekeeping

Corrective Safety Measures

a. Heat exhaustion or heatstroke during welding

- Keep the area well ventilated.
- Provide air blower if necessary.

- Provide drinking water and disposable cups near the welding area.
- First-aider with first-aid kit must be available at the working area during working hours.
- Consider work rotation during hot weather conditions.

b. Smoke produced from melting metals and welding rods

- Provide respirators for all workers involved in welding.
- Keep the area well ventilated.

c. Sparks and noise produced from grinding machines

- Provide earplugs or earmuffs for all workers at the fabrication shop and at or around the welding area.
- Provide safety glasses and welding shield protection for the welders.
- Provide fire blankets or a welding curtain around the working area.
- Fire extinguisher(s) must be available near the welding area.
- A trained fire watch must be available at site during welding activity.

d. Slips, trips, and falls

- Provide tin cans for welding rod butts.
- Minimize the length of the welding cables run around the area.
- Organize extension cords and welding cables to prevent tripping hazards.

e. Poor housekeeping

- Provide trash cans with lids to keep the garbage from blowing away.
- Perform housekeeping before and after working hours.

Heavy Equipment Operation

Potential Hazards and Conditions

a. Struck by or caught in between
b. Poor maintenance of equipment
c. Limited access going to the construction site

Corrective Safety Measures

a. Struck by or caught in between

- Install physical barricade around the area where the heavy equipment is in operation.
- Provide flagman to direct the operator to manoeuvre the equipment.
- Provide fluorescent vest to the flagman, who should always be in line of sight with the heavy equipment operator.
- The operator must be trained and possess a government-issued driving license.
- Install a backup alarm and beacon light in order for all to identify the location of the equipment.

b. Poor maintenance of equipment

- Carry out periodic maintenance of heavy equipment, and make sure it is in operational condition.
- Assign a full-time mechanic at the jobsite, and perform regular inspection.
- Perform daily inspection checklist before start of work and prior to moving the equipment from the motor pool.

c. Limited access going to the construction site

In order to simplify the hazards observations during the site walkthrough with the team participants, the data can be entered to the template in Appendix A.

Chapter 9

Developing a Method Statement

Have you heard about the method statement for the construction site that you are currently working at? If your answer is yes, then it is well and good for you. However, if this is the first time you have heard about it, here is your overdue opportunity to learn how to prepare it and perhaps increase your professional standing in the company.

A method statement is a structured process to address the sequence of the activities and methodology of executing a job to ensure quality and safety. It's an activity that is extracted from the execution plan in general. This process is a very effective approach to identify the project requirements to complete a critical activity within the project scope.

The advantage of developing a method statement is to inform construction companies whether they have resources to perform the job safely and meet the company owner's expectations. Upon reviewing the method statement, the company owner representative can understand the construction company's qualifications and assess whether the construction company knows the methodology to complete the job safely without interrupting the operation. In addition the method statement, construction companies will also prepare a structured job safety analysis (JSA) on each critical activity involved in the project (see chapter 10).

Who should prepare the method statement of each critical task in the project? Normally, every project has its own execution plan, for a government project, a private sector job, or an oil and gas producing facility. The method statement addresses the sequence of tasks that needs to be implemented in the project during the construction. Design engineering consultants, the project owner, and project engineer are familiar with the sequence of the project execution plan. Therefore, they must be involved in developing the method statement for the project.

The site supervisor and foreman, with the assistance of quality and safety staff, are essential parts of the team preparing the method statement so as to address the quality and safety standards in the project. The sequence of tasks in a critical activity should be defined clearly in the method statement to enable the workers to follow it when performing construction activities associated with each critical task.

Recommended best practices for developing a method statement to any critical task in the project suggest that it needs to address and define the following parameters in details. Accordingly, what is the purpose of the method statement apart from the execution plan of the project?

- Define the limitation of the project scope dealt with by the method statement.
- Identify the references, i.e., standard specifications, safety requirements, and drawings.
- Define the methodology and the sequence of critical tasks involved in the scope of work.
- Enumerate the general safety requirements and best safety practices of each activity.
- Attach the drawings that support the method statement for reference.
- Indicate the duration of work with manpower allocations and equipment in the method statement.

As soon as the document is consolidated and completed, it shall be submitted to the client project management support or design consultants for review and acceptance. Submission of these documents should be at least one week in advance prior to starting construction in order to give lead time for the reviewer to read through the sequence of the job. If the reviewer finds deficiencies in the method statement, the documents shall be returned to the contractor companies for revision. The client project management shall send a written notification to the construction companies as soon as the method statement meets the requirements. Upon receiving acceptance of the method statement from the client, the construction company may begin the construction activities.

With a detailed method statement for complex activities in the project, the construction company is able to foresee and prepare the resources to carry out the job safely and meet the client's expectations. In addition, the method statement helps prevent plant operation interruptions. A sample method statement provided in this book may help guide site management personnel who will be involved in developing a method statement for work identified as critical and more complex activities within the project scope.

This chapter will help the project engineer, site supervisor, and field staff develop a structured method statement for complex construction activities. The knowledge of individual in the project will learn a new strategy applies to method statement.

Purpose

The purpose of this method statement is to outline the procedure to carry out the job safely during the entire work of removing the existing 69 kV overhead line in Plant X and install the new 115 kV overhead transmission line from Plant X to Plant Y.

Scope

Removal of the existing 69 kV overhead line conductors mounted on the old wooden transmission poles from Sub-5 inside Plant X to a new transition yard. It includes the removal of conductors from the double dead end (DDE) pole of the suspension wooden pole. The double dead end pole is part of this statement to be removed. This shall only be executed when both the 69 kV and the new overhead transmission line outage are confirmed. Safety precautions shall be observed during the entire course of this activity.

References

- Permit to Work System
- Heavy Equipment Operator Testing and Certification
- Crane Lifts; Types and Procedures

Methodology

Part One

Sequence of work

1. After both lines are confirmed de-energized, mobilization of equipment shall commence.
2. The general arrangement of the old overhead line conductor has to be demolished. Install barricades with warning tapes around the area where the wooden pole is located. Flagmen

shall be provided to control the traffic on both sides of the road, if necessary.
3. At the suspension pole, guy wires shall be installed on the pole at the topmost wooden arm level with the use of a man lift as platform. It will be temporarily supported and restrained by loader equipment.
4. The conductor will be gripped and securely attached with the crane hook before cutting the conductor from the dead end pole.
5. When done, the man lift will move to the first suspension pole to cut the conductor.
6. When the line has been cut, the loader will move to the next pole and restrain the conductor using a guy wire.
7. This method will be repeated with the other lines in each pole up to the dead end pole on the opposite end.
8. All the conductors will be gradually lowered to avoid contact with the existing perimeter fence.

Part Two

1. After completing Part One of this method, the DDE wooden pole is now ready to be removed. Overhead conductor shall be removed the same way as described in Part One and secured on the existing DDE pole outside the fence.
2. The man lift shall assist the personnel in attaching and securing the web sling wrapped around the pole, starting from the bottom of the wooden pole with a double wrapped choker hitch and then attaching it to the main hook block or whip line hoist of a crane. The crane will then hoist slowly to tighten the web sling snugly in place.

3. As soon as the web slings are securely attached with the wooden pole, the crane will hold it continuously, while the workers cut the wooden pole at the bottom by using a chainsaw and slowly pull the wooden pole away. A tag line has to be secured on the pole while cutting or lifting is in progress.
4. Place the wooden pole in a safe area and further transport to designated dumping area.

General Safety Requirements

- Construction contractor shall provide a full-time site supervisor(s) who will be responsible to oversee this job, with full authority to give direction to the workers for carrying out the activity in a safe manner.
- Ensure that the crane operator has a government-issued license and valid certificate to operate the crane according to its type and model.
- Ensure the rigger has a valid rigging certificate.
- Work permit must be obtained prior to any activity in the area.
- Prior to lifting, the area must be well barricaded, and warning signs such as "lifting in progress" and "caution: falling objects" must be posted in the area.
- Safety personnel must be at the site during the entire course of activity. All other personnel not involved with the activity must be out of the lifting area.
- Prior to any lifting, a toolbox meeting must be conducted. All dangers during the course of the activity must be identified and addressed with safety precautions.

- Construction contractor shall ensure that all construction personnel shall have the necessary PPEs, including harness with shock absorbing lanyard. Safety officer shall inspect each harness and lanyard prior to use.
- Safety officer with the assistance of a rigger must inspect all lifting equipment such as sling and shackles. A softener pad must be installed at the pickup point of the steel member to prevent damage to the sling.
- A trained first aid worker and first aid kit must be at site.
- Construction contractor shall clear the area of materials such as concrete blocks, wood remnants, and steel after the erection is completed.
- Provide man basket at the jobsite, and ensure it is available when needed. The inspection sticker of the man basket must be valid, if required.

Chapter 10

Steps in Developing a Job Safety Analysis (JSA)

This chapter discusses the steps in developing a job safety analysis for a critical job in the project. It will support the project engineer and site supervisor, with the assistance of the safety officer, to develop a structured job safety analysis for critical activities. It addresses the sequence of critical tasks from the beginning until completion. The steps of each critical task should be provided by the project engineer and site supervisor, and the job safety analysis worksheet must list the potential hazards. The identified hazards in every task shall be provided with the recommendations.

Chapter Objectives

After reading this chapter, you should be able to do the following:

- Evaluate whether the job is to be considered as critical and requiring a job safety analysis.
- Enumerate the sequence of critical steps for each job.
- Identify the potential hazards of each step, and provide hazard controls or recommendations to mitigate the hazards.
- Enumerate the protective measures for each hazard involved with every work activity.
- Refer to the sample job safety analysis in this chapter, and be familiar with it.

Nowadays many construction companies are required to develop a job safety analysis for a certain activity in the project which is considered to be a critical activity. Some of the company's personnel are familiar with when a job safety analysis is required, especially those with experience working in oil and gas facility projects. Those working with a prime contractor are likely to deal directly with the client to provide safety requirements for the project, while other contractors' site management personnel – i.e., project engineer, site supervisor, and safety officer – are not yet familiar with job safety analysis requirements in the project. Perhaps their field experience has involved infrastructure projects rather than involvement with existing facilities operation. But why has job safety analysis been deemed important for this project? In fact, what is job safety analysis? And who will develop the job safety analysis for the project? These questions are common around the construction site and can delay the start of construction activity if this unforeseen requirement has not been submitted to the client representative ahead of time for review.

To answer these questions, you should know first the definition of job safety analysis as per Saudi Aramco Guide Number 06-003-2013, which states that Job Safety Analysis (JSA) is a structured method for developing task-specific procedures for critical jobs. A JSA breaks a job into basic steps; identifies potential hazards associated with each step; and recommends actions for each step to eliminate, control, or minimize hazards. (Since this requires identifying job steps and identifying the potential hazards, refer to chapter 8, Developing a Hazard Identification Plan, for guidance.) The purpose in developing the job safety analysis on every critical task is to ensure that the workers are protected from any potential incidents and consequent injuries that may occur when performing the job.

Why has job safety analysis been given priority for this project? It is simply because the tasks involved in the construction activity are considered as critical tasks and thus require identifying the sequence of basic steps for each task to prevent interruption of plant operations. Obviously, the project engineer, site supervisor, and foremen are responsible to oversee the project, and they will write down the basic steps of activities in the job safety analysis worksheet. Familiarity with the process of hazard identification will help the team members to develop the JSA and complete it in a timely manner.

The construction company shall organize a team and select the personnel who will participate in the site walk-through. The selected personnel should be trained on conducting the JSA. A site supervisor who has previous experience in developing JSAs could be a great help in identifying the potential hazards of critical jobs or equipment used in specific tasks. All tasks identified during the walk-through which are considered to be critical tasks shall be detailed and prioritized as per the job requirements.

The project engineer shall arrange the sequence of activities and break it down in the worksheet from start to finish of each critical task. The team member shall identify the potential hazard in every identified activity. Each hazard shall be addressed and provided with control measures or corrective actions.

A completed JSA worksheet shall be shared with workers during safety meetings and discussed before work starts. A copy of the worksheet shall be kept in the record and readily available for reference. See sample narrative job safety analysis for "Demolition of Existing emergency shutdown (ESD) Cabinets."

Sequence of Basic Steps

Step 1. Turn off the power supply to emergency shutdown (ESD) cabinets

Potential Hazards

- Plant operation interruptions
- Electrocution
- Injury

Hazard Controls

- Only qualified, trained, and authorized personnel will de-energize power supply to ESD cabinets.
- Perform power supply system inspection to make sure that the power supply to cabinet is correct.
- Install isolation lockout and tag-out to the main panel board.
- Use appropriate electrical rubber gloves with stamp indicating the voltage rating resistance.
- Use electrical aprons rated with voltage rating resistance, along with complete headgear and safety glass protection.
- Provide rubber mats when removing electrical power supply and instrument connection in the cabinet.

Step 2. Disconnect all the wiring and control cable in the cabinet

Potential Hazards

- Accidental start-up
- Electrocution
- Slip and struck by

Hazard controls

- Disconnect all the wiring connected from the cabinets to circuit breaker.
- Install lockout and tag-out all de-energized circuit breaker and controls.
- Hand tools should be in good condition and properly maintained.
- Perform housekeeping before the end of each workday.

Step 3. Demolish the old delta provox controller

Potential Hazards

- Caught in between
- Pinch point
- Struck by
- Injury
- System shutdown

Hazard Controls

- Only competent personnel are authorized to perform demolition work with close supervision.
- Verify the tagging number before removing the connection of controller.
- Organize the tools and materials around the working area.
- All electrical hand tools should be insulated and voltage rated.
- Perform housekeeping before the end of each workday.
- Ensure that all tools are in good condition.

Step 4. Remove the old nuts, bolts, and power supply frame

Potential Hazards

- Cuts
- Bruises
- Falling objects

Hazard Controls

- Wear gloves when removing bolts and nuts connected to the cabinets.
- Use safety helmet to protect head from falling objects during the demolition.
- Wear eye protection at all times when removing bolts and nuts.

Step 5. Remove existing cable from underneath the raised floor

Potential Hazards

- Trips and falls
- Pinch point
- Plant interruption
- Cable tray damage below the raise floor

Hazard Controls

- Install barricade around the tile opening.
- Use proper tile lifter on the raised floor.
- Do not stand on the existing cable trays underneath the raised floor.

- Use proper lifting technique for the raise floor tile.
- Provide protection for existing cable trays below the raised floor.

Step 6. Housekeeping and transporting a demolished cabinet

Potential Hazards

- Falls
- Struck by or caught in between

Hazard Controls

- Secure the cabinet at the back of the trailer truck prior to transport from the jobsite to designated storage area.
- Provide flagmen to direct the forklift when loading the cabinet.
- Heavy equipment such as forklift or crane must be fitted with backup alarm.
- Maintain continuous contact with the heavy equipment when loading the equipment.

Appendix A

Hazard Identification Plan (HIP) Form

S/N	Activities	Potential Hazards	Control Measures

Appendix B

Job Safety Analysis (JSA) Form

PLANT/EQUIPMENT		Department:	Date:
		Division:	JSA #:
Description of Activity:			
Prepared by:		Reviewed by:	Approved by:
		Date:	Date:
Safety Equipment:			
Steps	Sequence of Basic Job Steps	Potential Hazards	Hazards Controls

References

American Society for Testing and Materials (ASTM):

ASTM A500, Standard Specification for Cold-Formed Welded and Seamless Carbon Steel Structural Tubing in Round and Shapes

American National Standard Institute (ANSI):

ANSI Z359, Fall Protection

ANSI Z87.1, American National Standard for Occupational and Educational Personal Eye and Face Protection Devices

ANSI Z89.1, American National Standard for Industrial Head Protection

ANSI A92.5, Boom Supported Elevating Work Platforms

European Committee for Standardization, European Standard (EN):

EN 74-1, Couplers, Spigot Pins, and Baseplates for Use in Falsework and Scaffolds

EN 10219-1, Cold Formed Welded Structural Hollow Sections of Non-Alloy and Fine Grain Structural Steels

U.S. Code of Federal Regulations (CFR):

29 CFR 1910, Subpart I, Personal Protective Equipment

29 CFR 1910, Subpart O, Machinery

29 CFR 1926, Subpart I, Tools – Hand Tools

29 CFR 1926, Subpart L, Scaffolds

29 CFR 1926, Subpart P, Excavations

29 CFR 1926, Subpart X, Stairways and Ladders

2009 International Building Code

Special Thanks To The Following Persons Who Have Inspired Me To Write This Book

Mr. William "Robert" Crumpton, IV, CMIOSH, CSP

Engr. Alan K. D. Abellana, PME, Loss Prevention Engineer

Engr. Peter Q. Panganiban, Safety Advisor

Engr. Fernando M. Crisosto, Safety Advisor

Engr. Jesus G. Pedines Jr., Safety Advisor

Engr. Levi S. Alejo, Division Compliance Coordinator

Engr. Renato G. Barraca, Division Safety Coordinator

Engr. Lucresio L. Dagcutan, Safety Advisor

Engr. Ricky Salavia, Scheduler Engineer

Engr. Conrad A. Castillo, Site Superintendent

Engr. Isidro D. Pastrana. Site Superintendent

Engr. James Vagilidad, Safety Supervisor

Mr. Randy Dupaya, Safety Supervisor

Mr. John Villavicencio, Assistant Engineer

About the Author

Dr Marfa has more than twenty-five years of experience in the field of safety in oil and gas plant projects, monitoring multimillion-dollar projects of Saudi Aramco. A civil engineer by profession, he graduated from University of Cebu in 1985. He has held engineering jobs in a private construction firm, the Department of Public Works and Highways (DPWH), and the Department of Environmental and Natural Resources (DENR), Region 8, Philippines. Currently he is working with Hassan A. Karim Al-Gathani & Sons Company with SMP contract for Saudi Aramco projects, assigned in the Southern Area Oil Projects Division as Construction Engineer I, Safety Superintendent/Coordinator.

He holds a master's degree in business administration (MBA) from Philippine Christian University (PCU) administered by Al-Andalus International School, Al-Khobar extension program and graduated with a doctor of philosophy in business management in 2011. He is a Registered Safety Practitioners (RSP) and an Occupational Safety and Health Consultant (OSH Consultant) under DOLE-BWC Philippines. He is certified by OSHA for the construction train-the-trainer outreach program and authorized as a trainer in Construction Hazard Recognitions, Scaffolding Safety,

Crane and Rigging Safety, Confined Space Entry, administered by Saudi Aramco for continuing education through the University of California San Diego (UCSD) extension training program. He has also obtained a certification for Process Hazard Analysis (PHA) and Hazard Operability Study (HAZOP).

With a good background with the Saudi Aramco Construction Safety Manual General Instructions related to safety requirements, and familiar with engineering standards applicable in the construction industry, he has acquired authorization to provide safety training as an approved Professional Training Partner (PTP) through the International Safety Education Institute (ISEI), UC San Diego Extension.

He is incumbent chairman of the board of the non-profit Philippine Society of Safety Practitioners–Middle East Region (PSSP-MER), a former Philippine Society of Safety Engineers–Middle East Chapter (PSSE-MEC) president (2000–2002), and a lifetime member in the Philippine Institute of Civil Engineers (PICE), as well as former vice president of PICE Middle East Chapter (2005).

www.ingramcontent.com/pod-product-compliance
Lightning Source LLC
Chambersburg PA
CBHW030758180526
45163CB00003B/1080